Amsco's Science

Grade 6

Paul S. Cohen

Former Assistant Principal, Science
Franklin Delano Roosevelt High School
Brooklyn, New York

Anthony V. Sorrentino, D. Ed.

Former Director of Computer Services
Former Earth Science Teacher
Monroe-Woodbury Central School District
Central Valley, New York

Amsco School Publications, Inc.
315 Hudson Street / New York, NY 10013

The publisher wishes to acknowledge the helpful contributions of Melissa and Tom McFeely and Martin Solomon, who acted as consultants in the preparation of this book.
In addition, we wish to thank Jules J. Weisler for his valuable contributions.

Text Design: Howard Petlack, A Good Thing, Inc.
Cover Design: Meghan Shupe
Editor: Madalyn Stone
Composition: Publishing Synthesis, Ltd., New York
Artwork: Hadel Studio

Please visit our Web site at: *www.amscopub.com*

When ordering this book, please specify:
R 249 H or AMSCO'S SCIENCE Grade 6, *Hardbound*

ISBN-978-1-56765-922-1 (Hardbound Edition)

Printed in the United States of America
1 2 3 4 5 6 7 8 9 10   10 09 08 07 06

Contents

Introduction: The Nature of Scientific Inquiry

Unit

1

Simple and Complex Machines

Unit

2

Weather

Unit

3

Diversity of Life

Unit

4

Interdependence

To the Teacher

Amsco's Science: Grade 6, the first of a new, 3-book series for middle school science from Amsco School Publications, Inc., provides complete, clear, and concise coverage of the concepts taught in a yearlong 6th grade science class. As a primary text, *Amsco's Science: Grade 6* contains ample activities and exercises that will encourage the inquiry processes of science education. As a supplementary text, it can be used to reinforce conceptual understanding that more activity-based texts may lack.

The text consists of nine chapters divided into an introductory chapter and four units. Chapter 1, The Nature of Scientific Inquiry, introduces students to the processes, knowledge, and skills associated with science. Unit 1, Simple and Complex Machines, integrates the role and impact of energy and machines in our lives. Unit 2, Weather, discusses how matter and energy interact to produce weather patterns. Unit 3, Diversity of Life, shows how the transfer of matter and energy through biological communities supports the diversity of living things. Unit 4, Interdependence, discusses how interdependence is essential in maintaining life on Earth.

The text presents the major ideas of each topic in a clear manner that is easy to understand. It is aligned and corresponds with the National Standards (as well as those of New York City and New York State), and it can be used with any integrated science middle school curriculum. Each chapter is divided into lessons that allow teachers flexibility. You can teach the lessons in one day, or more, depending on the needs of your students. Conceptual understanding of topics is emphasized, and a balance of activities is presented to reinforce the topics.

Each chapter has an introductory page listing the contents of the chapter, a statement about what is in the chapter, a related career-planning feature, and a list of some Internet sites that provide additional knowledge or activities. Lesson titles are written in the form of a question and have clearly stated objectives. At the beginning of the lesson, a list of important vocabulary terms is presented—with a phonetic pronunciation guide—in the order in which they appear in the text. Each term is printed in **bold-faced** type in the lesson, and all of these terms are in the Glossary at the back of the book. Terms of lesser importance, or that have been introduced earlier, are in *italics*. Lessons consist of several topics that are visually explained with diagrams and pictures, as well as summarized in tables.

Each lesson is followed by Multiple Choice and Thinking and Analyzing Questions that reinforce the main points of the lesson. Many of the questions are taken directly from past exams. Activities and Skill Exercises related to the lesson provide additional reinforcement and an opportunity for students to do science at home or in school. Many of the activities are designed using easy to obtain materials commonly found around the home. Others require access to the Internet. Most lessons also contain a special boxed feature called "Interesting Facts About. . ." that presents little-known facts about the topic being discussed, or a related topic.

Finally, at the end of every chapter there are Review Questions. The questions are divided into four sections: Term Identification, Multiple Choice, Thinking and Analyzing, and a Crossword Puzzle. In the Term Identification questions, students are asked to match a word with its definition, and also to contrast it with another term. The Multiple Choice and Thinking and Analyzing questions are similar to published exam questions. The Thinking and Analyzing questions include short answer, diagram analysis, and reading comprehension questions. The Crossword Puzzle is a fun way of reviewing the important vocabulary of the chapter.

Chapter 1

Scientific Inquiry

Contents

Scientists make careful observations.

What Is This Chapter About?

Welcome to the world of science. You may be wondering what you are going to find in this world, and in this textbook. This text includes chapters on physics, chemistry, weather, and ecology. Why are these different topics all considered science? Scientists in these areas obtain and organize information in much the same way. They follow procedures, known as the scientific method, to make predictions, collect data, and form theories. These theories attempt to explain some part of the natural world.

In this chapter you will learn:

1. What different fields of science have in common.

2. How scientists look at the world, make observations, and form theories.

3. How to design an experiment and present the results using graphs and tables, and how to use models to explain difficult concepts.

Career Planning:
So, You Want to Be a Scientist

What kind of a scientist would you like to be? Science is broken down into many different subjects, ranging from astronomy to zoology. Research scientists discover new information used to develop new products, while quality control scientists test products for quality and purity. Some scientists become science teachers. This field is called *science education*. The career opportunities in science are as varied as science itself. We hope that as you study each of the units of this book you imagine yourself as a scientist working in that field.

Internet Sites:

http://www.biology4kids.com/files/studies_scimethod.html Read about the scientific method and scientific proof.

http://www.nceas.ucsb.edu/nceas-web/kids/experiments/data/data.html Explore the basics of data and data tables and how to use different types of graphs.

1.1

What Do Scientists Do?

Objectives

Identify different areas of scientific study.

Identify science as a method of solving problems.

Terms

inquiry: a system of scientific investigation

data: information

observation: anything we perceive through one or more of our senses

research: gathering information, facts, and the opinions of others

experiment: an organized procedure to test an idea or gather new information

What Is Science?

Many dictionaries describe science as a body of knowledge systematically obtained. All branches of science obtain this knowledge through a system of investigation, called **inquiry.** This system includes making observations, posing questions, examining references, planning experiments, gathering and interpreting information, proposing explanations, and making predictions. Does that seem like a lot? With so much to do, it is no wonder that most scientists focus on just one particular area of science.

Science is often broken into three major categories: physical science, Earth science, and life science. Physical science includes the fields of chemistry and physics. Earth science includes astronomy, meteorology, and geology. Life science includes biology and ecology.

What does a biologist study? The word "biology" combines the prefix "bio," meaning life, with the suffix "logy," meaning "study." A biologist might study the structure of a cell, or the structure of a tree. Biology includes details about every living thing. The field of biology is so enormous that it must be broken down into smaller areas of study. Some of these include:

botany — the study of plants

zoology — the study of animals

cytology — the study of cells

Table 1.1-1 shows some fields or branches of physics, chemistry, geology, and biology.

You can see that science contains many different areas of study. All of these, however, use similar methods of investigation.

Table 1.1-1. Some Fields of Science

Physics	Chemistry	Geology	Biology
Optics – the study of light	Organic chemistry – the study of carbon compounds	Seismology – the study of earthquakes	Entomology – the study of insects
Acoustics – the study of sound	Metallurgy – the study of metals	Volcanology – the study of volcanoes	Parasitology – the study of parasites
Thermodynamics – the study of heat and its relationships to other forms of energy	Pharmacology – the study of medicine	Mineralogy – the study of minerals	Genetics – the study of heredity

CSI – Common Scientific Investigations

Have you ever watched a detective program on television? Scientists, like detectives, gather information in order to solve a particular problem, or explain a particular event. The detective gathers pieces of information, called "clues," while the scientist gathers pieces of information, called **data**. In both of these cases, the greater the amount of information gathered, the more reliable the conclusion.

Making Observations

Science begins with **observations**. An observation is anything we perceive through one of our five senses: taste, smell, sight, touch, and hearing. We use tools to improve the ability of our senses. For example, our sense of sight does not permit us to see small things, such as germs, nor far away objects, such as distant galaxies. Tools such as microscopes and telescopes extend our sense of sight.

Other tools can detect things that we cannot. Physicists use galvanometers (GAHL-vuh-NAH-miht-ers) to measure tiny amounts of electricity. A geologist may use a compass to detect Earth's magnetic field. A meteorologist (MEET-ee-uhr-AHL-uh-jist) uses a barometer (buh-RAH-miht-er) to measure changes in air pressure. The tools of the scientist greatly improve our ability to observe the natural world. (See Figure 1.1-1.)

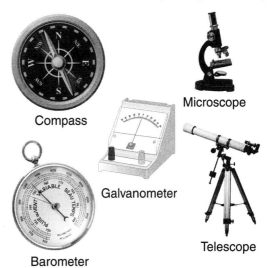

Compass

Microscope

Galvanometer

Telescope

Barometer

Figure 1.1-1. Some tools used by scientists.

Using Observations

What does a scientist do with the information obtained through careful observation? Like a detective, a scientist is a problem solver. To solve problems in science, a scientist first finds out what other scientists have already discovered. This gathering of facts, data, and the opinions of others is called **research**. From this research, solutions are proposed and tested. These tests, called **experiments**, provide more data. With enough data, the scientist can find answers to questions about the world we live in.

So, what do scientists do? What types of problems do they solve? These are difficult questions to answer because there are so many possibilities! Here is a short list of some of the many things that scientists do:

- Invent new medicines
- Find new fuels to make electricity
- Create new batteries for electronic devices
- Make metals stronger and lighter
- Find causes and cures for diseases
- Protect wildlife

- Improve the environment
- Solve crimes
- Discover new stars
- Forecast the weather
- Predict tsunamis
- Make buildings safer against earthquake damage
- Predict volcanic eruptions
- Devise better sound systems
- Develop devices to improve sight and hearing
- Improve gas mileage in cars
- Make food taste better
- Protect crops from pests and diseases
- Develop bulletproof materials to protect police and soldiers
- Develop more powerful weapons
- Keep clothing from staining or wrinkling

Can you think of any others? In fact, the list could go on and on. Science is not just *what* we do, but *how* we do it. There are certain methods that are common to all areas of science. The rest of this Chapter will explore many of these methods.

Activity

Sometimes scientists need tools that help them see things better. If the objects they are studying are very far away, they need telescopes. If the objects they are studying are very small, they need microscopes. At the *Powers of 10* Web site, look at the world as it is seen from millions of meters away, and then move closer and closer until you are looking at a plant cell magnified millions of times. Visit the Web site at

http://micro.magnet.fsu.edu/primer/java/scienceopticsu/powersof10

to take this amazing journey.

    1. Which distance did you find most interesting?

    2. What did you observe at this distance?

    3. If you were a scientist, what could you be studying at this distance?

Interesting Facts About Science

Aristotle, one of the greatest philosophers of the ancient world, placed great importance in studying the natural world. In 335 BC, Aristotle began a school of organized scientific studies. He defined scientific knowledge and why it was important. In other words, he single-handedly invented science. Yet, Aristotle was actually anti-science by today's definition.

Aristotle felt that to understand the natural world we have to observe it. According to Aristotle, if we do an experiment, we are controlling or changing the natural world, and it isn't natural any more. Fortunately, modern scientists do not agree! Nearly all of the modern advances in science are the result of controlled experiments.

Although science has been studied for thousands of years, most modern ideas of science were not developed until the end of the middle ages. Sir Francis Bacon, a philosopher in the early 1600s, said that the way to master the natural world is through scientific inquiry. Bacon was the originator of the expression, "Knowledge is power."

Questions

1. Which of the following terms includes the other three?
   (1) physics   (3) biology
   (2) chemistry   (4) science

2. The sum of all the methods that scientists use to study the natural world is called
   (1) scientific inquiry
   (2) scientific experimentation
   (3) research
   (4) observation

Thinking and Analyzing

The diagram at right is called a tree. Actually, it looks more like an upside-down tree. The root of the tree is at the top, and the leaves are at the bottom. A tree shows how items are related to one another. It helps us sort, or categorize (place into categories) connected ideas. The tree shown here begins to sort out the fields of science.

   1. Complete the tree by filling in the blank spaces. Use examples from this lesson to help you.

   2. Extend the tree by adding six more leaves.

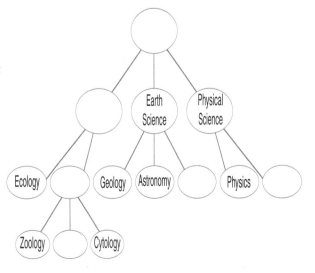

1.2 How Do Scientists Solve Problems?

Objectives
Describe the steps used in scientific inquiry.
Make hypotheses based on observations.
Describe how hypotheses are tested.

Terms
problem: the question to be investigated

hypothesis (hy-PAH-thuh-sis): a possible answer to a problem, based on observation and research

scientific method: an organized, step-by-step approach to problem solving

You have learned that scientific inquiry begins with observations, and that scientists use these observations to solve problems. Scientific inquiry, however, is a useful tool for everyone, not just for scientists. Let's see how you might use the methods of science to solve an everyday problem.

Everyday Science

Suppose you walk into a room and notice an unpleasant odor. You say, "It stinks in here!" You have just made an *observation*. "What is causing that horrible smell?" You have just stated a **problem**. A problem is always stated in the form of a question. You might now say, "Something must have died in here." You have now made a guess as to the solution to your problem. You observed a foul odor, and you suspected that the cause of the odor was a dead animal. A

possible answer to a problem, based on observations or prior knowledge, is called a **hypothesis.** (Hypotheses (hy-PAH-thuh-seze) is the plural of hypothesis). If you had never smelled a dead animal before, you would not have been able to guess the cause of the odor. The best hypotheses are those that are based on a large number of careful observations. Once a hypothesis has been posed, it is generally tested through further observations and experimentation. In the case above, you would want to find the dead animal before you conclude that your hypothesis is correct. This organized, step-by-step approach to problem solving in science is known as the **scientific method**.

Here is another everyday example of the scientific method at work.

On the first warm day in March, Jennifer takes out her bicycle, which she has not ridden all winter. She finds it very difficult to pedal. She also is not able to ride as fast as she used to, even though the wind is behind her. Jennifer has just made two observations. What problem does she wish to solve? She might wonder, "Why can't I ride as fast as I used to?" Recall that a hypothesis is a possible answer to a problem. She could form a hypothesis at this point, such as, "I am really out of shape," or she could make some additional observations. She might wish to turn the bicycle over and try to turn the pedals. They seem to be more difficult to turn than she remembers, but she doesn't know why. At this point, she might use another important scientific technique — **research**. Where should Jennifer look for the information she needs? She could use the Internet, or a library, but it is likely that she would first consult the manual that came with the bicycle. In the manual,

she finds a section called "Caring for Your Bicycle." In that section, it says that you should oil the bicycle regularly.

This research causes her to change her hypothesis. Her new hypothesis is that she cannot ride as fast as she used to because the bicycle needs to be oiled. How can she test this hypothesis? She oils the wheel, and compares the way the wheel turns before and after she oils it. It appears to turn more easily after she oils it. Another ride on the bicycle could prove her new hypothesis correct.

Jennifer used the *scientific method*. She made some *observations*, and *posed a problem*. She did *research*, formed a *hypothesis*, and did an *experiment* to test that hypothesis. On the basis of her observations, her research, and her experiments, she drew a *conclusion*. Jennifer followed all of the steps of the scientific method, in the correct order. Sometimes, we are all scientists.

Questions

Use the following information to answer questions 1–3.

Many scientists believe that Earth's climate is getting warmer because of an increase in the amount of carbon dioxide in the air.

1. A scientist wishes to investigate the change in Earth's climate. He might state the problem as
   (1) Earth's climate is getting warmer.
   (2) The amount of carbon dioxide in the air is increasing.
   (3) Why is Earth getting warmer?
   (4) An increase in the amount of carbon dioxide is causing Earth to get warmer.

2. Which of the following is a possible hypothesis made by a scientist studying this problem?
   (1) Earth's climate is getting warmer.
   (2) The amount of carbon dioxide in the air is increasing.
   (3) Why is Earth getting warmer?
   (4) An increase in the amount of carbon dioxide is causing Earth to get warmer.

3. The scientist investigates the average temperature readings listed in 20 different regions of the world over the past 50 years. This activity is best described as
   (1) research
   (2) forming a hypothesis
   (3) experimentation
   (4) drawing conclusions

Thinking and Analyzing

1. Place the following steps in the correct order from the first step in the scientific method to the last:

   - Research
   - State a hypothesis
   - Pose a problem
   - Make an observation
   - Design an experiment

   Base your answers to questions 2–4 on the following pararagraph.

   A poultry farmer was worried because his turkeys were underweight. The farmer

Interesting Facts About Scientific Observations

How one observation changed the world

Bacteria are tiny organisms that can be seen by using a microscope. Some bacteria cause serious diseases, such as strep throat and tuberculosis. While studying a culture of bacteria in 1928, the Scottish bacteriologist (back-TEER-ee-AHL-uh-jist) Alexander Fleming noticed that there were no bacteria growing in one region of a particular sample. (See Figure 1.2-1.) By observing this sample more carefully, he discovered that a mold was growing in the region where there were no bacteria. Fleming might have hypothesized at this point that the mold kills bacteria. However, it is important to base a hypothesis on a large number of observations. The greater the number of observations, the more likely the hypothesis is correct. On the basis of just one experiment, Fleming could not be sure that it was the mold, and not some other unknown factor, that had killed the bacteria. In this instance, Fleming went on to test this hypothesis by performing hundreds of experiments. He

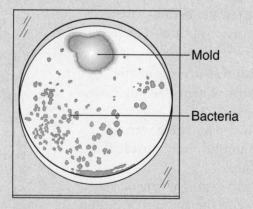

Figure 1.2-1.   Fleming observed that bacteria do not grow near a mold.

had noticed a fox near his farm, which was upsetting the turkeys. He trapped and relocated the fox, but his turkeys continued to be underweight. Eventually, he noticed an increase in the number of mice in the area where the turkeys were kept. The farmer left some of the turkeys' food just outside of the turkeys' reach. The food disappeared overnight.

2. State the problem being investigated by the farmer.

3. What was the original hypothesis formed by the farmer?

4. How would you revise that hypothesis based on his observation of the mice?

5. What should the farmer do to help increase the weight of his turkeys?

confirmed that it was a chemical in the mold that was killing the bacteria. The mold was called penicillium, and the chemical became known as penicillin.

Fleming published the results of his experiments in a scientific journal. His work was available to other scientists who were searching for ways to fight disease. Publishing and sharing your findings is an important part of the scientific method, as is investigating the work of other scientists. Today's scientists have a wide variety of research sources available to them. These include scientific journals (which publish the results of experiments), books, periodicals, and the news media. Nearly all of these are available on the Internet and in libraries.

In 1938, ten years after Fleming's research was published, other scientists realized that Fleming's work could have important applications. Howard Florey, an Australian scientist, was doing research at Oxford University in England. Florey hired Ernst Chain, who had fled to England to escape Hitler's Germany. Chain read Fleming's paper, and began testing penicillin on mice. His controlled experiments showed that penicillin could cure bacterial infections. It was then tested on humans, and used with great success during World War II.

Fleming's careful observations, Florey's and Chain's research, and the work of many other scientists eventually led to the use of penicillin, a drug that has saved many millions of lives. Fleming, Florey, and Chain won the Nobel Prize in 1945 for the discovery of penicillin. For more information on Fleming and the discovery of penicillin, visit *http://www.pbs.org/wgbh/aso/databank/entries/ dm28pe.html*.

1.3 How Do Scientists Design Experiments?

Objectives

Design a controlled experiment.

Identify the dependent and independent variables in an experiment.

Identify the factors that determine the reliability of an experiment.

Terms

independent variable: a condition in an experiment that is deliberately changed by the scientist

dependent variable: the condition in an experiment that changes due to the effects of the independent variable

variable: a condition that changes during an experiment

control or **control group:** a standard to which the experimental group is compared

experimental group: a group in an experiment that receives the independent variable

Has anyone ever given you roses? The florist usually includes a little package containing a white powder. The powder is added to the water into which the flowers are placed. The roses are supposed to last longer when the powder is added. Does it work? Some people add aspirin to the water, while others add a penny. How can you determine the best conditions for maintaining cut roses? You can design and perform an experiment. Many experiments measure the effect of a single change in conditions. The condition that you manipulate (change) is called the **independent variable**. The condition that responds as a *result* of that change is called the **dependent variable**. If you designed an

experiment to measure the effect of a substance on the freshness of a cut flower, the independent variable is the substance that you add. The dependent variable is the length of time until the flower droops, or the petals drop.

Variables

In performing valid experiments, it is important to limit the number of variables. **Variables** are the conditions that change and affect the outcome of an experiment. In our flower experiment, we need to be sure that all the flowers we test are cut at the same time. We want to use only one type of flower. We need to keep the flowers at the same temperature, and have the same amount of light. The only thing that should vary in this experiment is the material that we add to the water. Anything else that might affect the flowers must be kept constant.

Controlled Experiments

Let's say that we perform our flower experiment on three identical flowers, placed at the same time in three glasses of water. One glass contains aspirin, one contains a penny, and the third contains the florist's powder. After three days, we notice that all of our roses have begun to droop. What conclusion can we reach? Did all three methods help equally in maintaining the roses? Perhaps, but does this mean that we should always use one of these methods when we get cut flowers? We can't tell! To find out whether *any* of these methods is effective, we need to include in our experiment a flower in water that contained *none* of our chemicals. By comparing the untreated sample to the treated samples, we can determine whether the treatment has any effect. The untreated samples are called the **control** or **control group**, while the treated samples are called the **experimental groups**.

In his experiments with penicillin, Fleming had to deal with such variables as food supply, light, temperature, and type of bacteria, any of which might affect bacterial growth. To limit these variables, he took two identical samples of bacteria under identical conditions, and then added penicillin to one of them but not to the other. In this way, he could confidently conclude that it was the penicillin that was affecting the

bacteria, and not some other variable.

An experiment that tests the effect of just one variable is called a *controlled experiment*. In Fleming's experiment, one group of bacteria was treated with penicillin, while another was not. The samples that were not treated with penicillin were the control group. The samples that *were* treated were the *experimental groups*. Only by comparing an experimental sample to a control can we draw valid conclusions.

Sample Size

In the flower experiment, we placed one flower in each of three mixtures, plus one, the control, in untreated water. Do you think it would make any difference if, instead of using just one flower in each group, we used several? When you receive a bunch of roses, no matter how you care for them, some of them droop before others do. There are individual differences that you cannot control. By using several flowers in each group and comparing the *average* result, you minimize the effect of individual differences. In general, the larger the sample size, the more reliable are the results.

After one at-bat, some baseball players are batting 1.000, while others are batting 0. Would you want to base your fantasy baseball team on statistics from the first at bat of the season? Would you base your decisions on the first game of the season? In baseball, as in science, conclusions are best made after a large number of observations are recorded.

A reliable experiment should produce the same result when it is repeated. The best way to make sure that the results are reliable is to perform the experiment several times.

Activity

Midge notices that when a raw potato slice is left out in the air it turns dark. She also observes that the potatoes in her potato salad never turn dark. She looks up several recipes for potato salad and sees that vinegar is used in all of them. Midge hypothesizes that vinegar prevents the darkening of a potato. You can test her hypothesis by performing the following experiment:

Materials:

4 small dishes or bowl	4 slices of raw potato
1 Tablespoon	1 Fork
White Vinegar	Water

Procedure:

1) To one dish add 2 tablespoons of water.
2) To one dish add 1 tablespoon of water and 1 tablespoon of vinegar and stir gently.
3) To one dish add 2 tablespoons of vinegar.
4) Add a slice of raw potato to each of the four dishes.
5) Wait about a minute and then turn over the slices of raw potato with the fork.
6) Check the experiment in one hour.
7) Let the experiment stand overnight and check again in the morning.
8) Record your observations

Answer the following questions:

1. Which potato darkened the most?
2. Which potato changed the least?
3. State the problem that was investigated in this experiment.
4. Identify the independent variable.
5. Identify the dependent variable.
6. Did your results agree with the hypothesis stated above?
7. Why was one slice of potato placed in water?

Midge's mother says that the reason the potatoes in the potato salad do not turn black is that they are **cooked** potatoes. She thinks that only raw potatoes turn black. Design an experiment to test Midge's mother's theory.

*Interesting
Facts About
Control
Groups*

Poliomyelitis (POH-lee-oh-mhy-uh-LIE-tis), also known as polio, was a devastating disease that once threatened all Americans, especially children. Today, there are no cases of polio in the United States thanks to a vaccine developed by Jonas Salk in the 1950s.

Salk first tested the vaccine on monkeys and then on small groups of volunteers to make sure it was safe. Then, in 1954, a massive experiment was designed to test the effectiveness of the vaccine on people. Almost 2 million children participated. This experiment was a double blind experiment, the most reliable type of test possible. In a *double blind experiment*, neither the subject nor the experimenter knows which subjects are in the experimental group and which are in the control group. Both groups were given injections of identical-looking liquids. The experimental group received the vaccine, while the control group received a harmless substance, called a *placebo*. The placebo looked just like the actual vaccine. Neither the doctors nor the children could tell which group the children were in. A placebo is used to keep the experimental group and the control group under exactly the same conditions.

The experiment proved that the vaccine was both safe and effective. Because both the experimental group and the control group were so large, the results of the experiment were extremely reliable. The following year, the Salk vaccine was given to children throughout the United States. Today, polio is a problem only in parts of the world where children are not vaccinated. There have been no cases of polio in the United States for at least a decade (10 years).

Questions

1. The most reliable experiments are those that have a
   (1) small sample size and many variables
   (3) large sample size and many variables
   (2) small sample size and few variables
   (4) large sample size and few variables

2. A gardener wishes to determine the effect that changing the amount of watering has on his tomato plants. Which is the independent variable in his experiment?
   (1) the amount of sunlight
   (2) the size of the tomatoes
   (3) the height of the plants
   (4) the amount of water

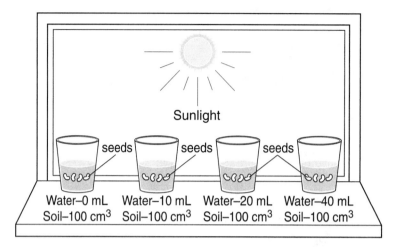

Use the experiment described below to answer questions 3 and 4.

A student set up the experiment shown in the illustration to learn about plant growth. The student added a different amount of water to four identical containers, each containing four seeds in 100 cubic centimeters of soil. All of the containers were placed in the same sunny location.

3. Which variables need to be held constant to make this a valid experiment?
   (1) amount of water and amount of sunlight
   (2) amount of sunlight and temperature
   (3) temperature and amount of water
   (4) amount of soil and the height of the plants

4. All of the following would make the results more reliable **except**
   (1) using different types of plants
   (2) using a greater number of seeds
   (3) maintaining the same temperature for all groups
   (4) repeating the experiment several times

5. In an experiment to answer the problem, "What is the effect of A on B?" which of the following statements is **true**?
   (1) A is the dependent (responding) variable.
   (2) B is the dependent (responding) variable.
   (3) Both A and B are dependent (responding) variables.
   (4) Neither A nor B is a dependent (responding) variable.

Thinking and Analyzing

1. A student goes skateboarding a few times a week. The student notices that she can go faster while skating on some level surfaces than on others. She hypothesizes that speed has something to do with the surface she is skating on. The student wants to design an experiment to test this hypothesis.
   a. Identify the independent (manipulated) variable in the experiment.
   b. Identify the dependent (responding) variable in the experiment.
   c. Identify *two* factors that will need to be held constant in the experiment.

2. Sharon wishes to investigate whether a certain brand of fertilizer really produces larger flowers. She grows petunias, pansies, and marigolds, adding the same amount of fertilizer to each plant at the same time. The plants are grown under the same conditions of soil, light, and temperature. When the plants blossom, Sharon measures the size of the flowers. She finds that the petunias are larger than the pansies or marigolds.
   a. Pose the problem that Sharon was attempting to investigate.
   b. What is the major error in the design of this experiment?
   c. Sharon concluded that the fertilizer works best on petunias. Is her conclusion valid? Explain.

1.4 How Do Scientists Present Results?

Objectives

Draw and interpret line graphs.

Draw and interpret bar graphs.

Interpret pie charts.

Terms

line graph: a graph that connects numerical data points

***x*-axis:** the horizontal axis of a graph on which the independent variable is shown

***y*-axis:** the vertical axis of a graph on which the dependent variable is shown

bar graph: a graph that uses length to show differences between experimental groups

quantitative (KWAN-tih-tay-tiv): involving measured values

pie chart: a graph that compares the parts of a whole by representing them as slices in a pie

Organizing Observations

When we wish to form a hypothesis based on a large number of observations, it is important to organize our observations in a logical fashion. Suppose that we are studying a chemical reaction occurring at various temperatures. The chemical hydrogen peroxide (H_2O_2) breaks down to produce water (H_2O) and oxygen gas (O_2). We can measure how long it takes to produce 50 milliliters (mL) of oxygen at five different temperatures. In the first test, performed at 30°C, it takes 8.0 minutes. In the second test, performed at 50°C, it takes 2.1 minutes. In the third test, performed at 10°C, it takes 33.0 minutes. In the fourth

test, performed at 20°C, it takes 16.0 minutes. Finally, in the fifth and last test, performed at 40°C, it takes 4.1 minutes.

Before we can draw a conclusion about the effect of temperature on this reaction, we need to organize our data in a more logical fashion. Table 1.4-1 shows two possible ways (*A* and *B*) we can organize the data.

Which table is more helpful to us in forming a hypothesis about the effect of temperature on the time needed to produce a given amount of oxygen? Table *A* has more data, because it includes the order in which the observations were made. However, the extra data are neither helpful nor important! By listing the results in order of increasing temperature, as in Table *B*, we make it easier

Table 1.4-1. Time Needed to Collect 50 mL of Oxygen from Identical Solutions of Hydrogen Peroxide

Table A			Table B	
Trial	Temp. (°C)	Time (minutes)	Temp. (°C)	Time (minutes)
1	30	8.0	10	33.0
2	50	2.1	20	16.0
3	10	33.0	30	8.0
4	20	16.0	40	4.1
5	40	4.1	50	2.1

to observe a trend, and easier to draw a conclusion. Table *B* shows a better way of organizing the data.

Using Table *B*, we can easily see that as temperature increases, less time is needed to collect 50 mL of oxygen. In other words, the higher the temperature, the faster the reaction.

The two columns in our table, temperature and time, represent our independent and our dependent variables. The independent variable is written first in most data tables. The temperature is the independent variable and the time is the dependent variable.

In a scientific table, each column has a heading showing the name of the variable. If the variable is a measurement, such as time or temperature, the unit is shown in parenthesis. In Table 1.4-1 *B*, the top row reads "Temp. (°C)" in the first column and "Time (minutes)" in the second column.

There are further conclusions that we can draw from Table *B*. Although we did not measure it directly, we can determine how long it would take to collect 50 mL of oxygen at 25°C. We predict that it would take less than the 16 minutes needed at

20°C, but more than the 8 minutes needed at 30°C. We might even guess that it would take 12 minutes, which is halfway between 8 and 16. However, we can make a more accurate guess by using an important visual tool for organizing data, called a **line graph**.

Using Line Graphs

In Figure 1.4-1 on page 18, we have graphed the data given in Table *B*. We chose to plot temperature on the *x-axis* (the horizontal axis) and time on the *y-axis* (the vertical axis). We then entered our five measured points, and connected the points. The lines connecting the points are *not* straight lines. Our five points do not lie on a straight line, so we draw a smooth *curve* to connect the points.

Line graphs are used to show the relationship between two numerical variables, the independent variable and the dependent variable. The *independent variable* is placed on the x-axis and the *dependent variable* is plotted on the y-axis. In our experiment, time is the dependent variable. The amount of time (to collect 50 mL of oxygen) depends on the chosen temperature.

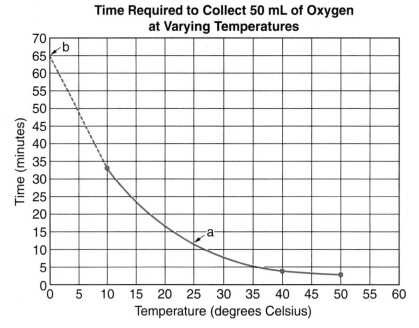

Figure 1.4-1. The line graph illustrates the effect of temperature on a rate of reaction.

Our five measured points show our five observations. A line graph enables us to predict points that lie between our measurements. From our graph we can conclude that, at a temperature of 25°C, it should take about 11 minutes to collect 50 mL of oxygen (see point "*a*" on the graph). All of our measurements were made between 10°C and 50°C. We can extend the curve to show what might happen at a temperature of 0°C. By extending the graph, we can predict that it would take about 64 minutes to collect 50 mL of oxygen at a temperature of 0°C (see point "*b*" on the graph).

By graphing our observations, we are able to predict values that lie between and beyond our measured points. More reliable predictions can be made if there are a greater number of observations.

Using Bar Graphs

Bar graphs are also used to help organize and illustrate observations. Suppose we want to compare the weather in New York State to the weather in some other areas of the United States. One area of interest might be the amount of rainfall. Table 1.4-2 shows the average annual rainfall in eight cities in the United States, three of which are in New York State.

A bar graph can be used to compare these eight observations, as shown in Figure 1.4-2. Looking at the bar graph, you can quickly observe that of the eight cities, Miami has the most rainfall each year, while Las Vegas has the least. Bar graphs are used to make clear and dramatic comparisons. Notice that in a line graph (Figure 1.4-1), both the *x*-axis and the *y*-axis are **quantitative** (represent measurements),

Table 1.4-2. Average Annual Rainfall in Eight Cities in the United States

City	Average Annual Rainfall (cm)
New York, New York	109
Albany, New York	91
Buffalo, New York	95
Cleveland, Ohio	90
Honolulu, Hawaii	60
Miami, Florida	146
Las Vegas, Nevada	11
Los Angeles, California	31

whereas in a bar graph (Figure 1.4-2), only the *y*-axis is quantitative.

Pie Charts

What part of the day do you spend sleeping? Suppose you average 8 hours of sleep per day. You spend 6 hours in school and

2 hours doing homework, 1 hour eating, and 1 hour exercising. The remainder of the day is free for other activities, such as watching television, listening to music, and surfing the Web.

To better understand how you spend your day, you can organize the data into a table. Table 1.4-3 on page 20 lists the various activities and the number of hours you spend in each. How can we represent this data graphically? You can use a bar graph to compare these time measurements, but since each time measurement is a part or percentage of the entire day, a better graphical organizer is a **pie chart**. Figure 1.4-3 on page 20 compares a pie chart and a bar graph of your daily activities. You can easily see from either graph that sleeping takes up more of your day than any other activity. You can also see from the pie chart that you spend a fourth of your day in school. A pie chart is the best way to represent the parts of a whole.

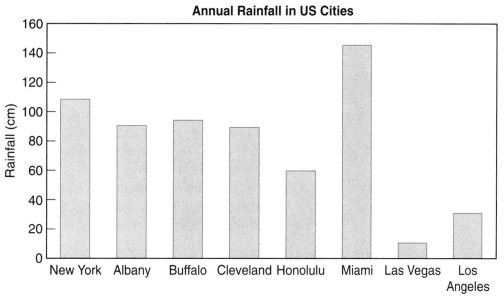

Figure 1.4-2. The bar graph compares the average annual rainfall in several US cities.

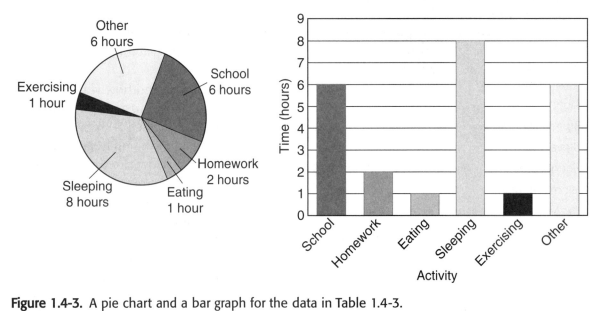

Figure 1.4-3. A pie chart and a bar graph for the data in Table 1.4-3.

Table 1.4-3 Time Spent in Various Activities

Activity	Time Spent (hours)
School	6
Homework	2
Eating	1
Sleeping	8
Exercising	1
Other	6

Interesting Facts About Graphs

Many computer programs distinguish between a line graph and an x-y plot. The graphs we have been discussing in the chapter are actually x-y plots. The line graph in a computer program is quantitative on the y-axis only. For example, if you want to plot the average temperature in New York City month by month, your x-axis is the month, while your y-axis is the temperature. Only the temperature is quantitative. In science, however, graphs are used to make quantitative predictions, and it is usually required that both the x-axis and y-axis be quantitative.

When using a computer program to make line graphs, many students select "Line Graph" rather than "X-Y Plot" and do the graph incorrectly.

Activity

You can create graphs online at *http://nces.ed.gov/nceskids/graphing*. Just select the type of graph you want, go to the data tab, enter the graph title, *x*- and *y*-axes labels, select the number of items (points) you want to graph, enter your data and labels, and view your graph.

An experiment was done to determine the relationship between the speed of a car and how far it travels before it stops after the brakes are applied (stopping distance.) The average results are listed below. Use the *http://nces.ed.gov/nceskids/createagraph/* Web site or another graphing program to create an *x-y* plot of this data. Print the graph. Now create a line graph from the same information. Print that graph, too.

Speed (km/h)	Stopping distance (m)
32	6
48	14
80	38
96	55
112	75

1. Compare the *x*-axis (horizontal axis) for each graph. How do they differ?

2. Which graph is a better representation of the relationship between stopping distance and speed?

Questions

1. In an experiment to determine if the height of a plant is affected if the plant is watered with different amounts of water, the height of the plant is the
   (1) independent variable and placed on the *y*-axis
   (2) independent variable and placed on the *x*-axis
   (3) dependent variable and placed on the *x*-axis
   (4) dependent variable and placed on the *y*-axis

2. Which of the following is the best way to graphically represent the average test scores of sixth graders, seventh graders, and eighth graders?
   (1) a bar graph   (3) a pie chart
   (2) a line graph   (4) a table

3. According to the bar graph shown in Figure 1.4-2 on page 19, which city receives less rainfall than Cleveland but more than Los Angeles?
   (1) New York   (3) Las Vegas
   (2) Miami   (4) Honolulu

4. The pie chart below compares the amounts of energy from different sources used in the United States each year.

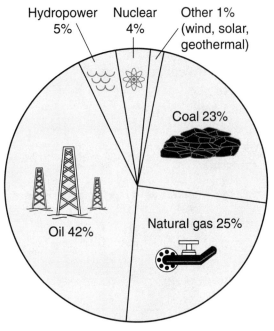

Hydropower 5%
Nuclear 4%
Other 1% (wind, solar, geothermal)
Coal 23%
Oil 42%
Natural gas 25%

Which two energy sources together provide more than half of the energy needs of the United States?
(1) nuclear and natural gas
(2) hydropower and oil
(3) oil and coal
(4) natural gas and coal

5. The bar graph below shows the number of cases of Lyme disease reported in the United States from 1983 to 1989. In which year was the reported number of cases less than it was the year before?

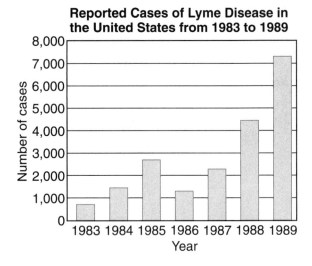

Reported Cases of Lyme Disease in the United States from 1983 to 1989

Number of cases

Year

(1) 1984   (3) 1986
(2) 1985   (4) 1987

Thinking and Analyzing

Base your answers to questions 1–6 on the information and diagrams below.

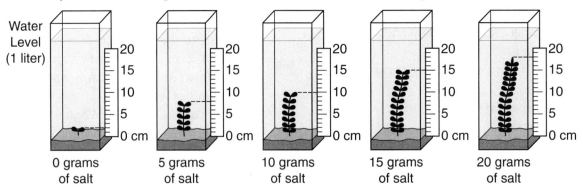

Water Level (1 liter)

0 grams of salt

5 grams of salt

10 grams of salt

15 grams of salt

20 grams of salt

Saltwater plants of the same species were grown in soil in separate containers with 1 liter of water. All of the plants were the same height at the beginning of the experiment. Different amounts of salt were dissolved in each container as shown in the diagrams. All other conditions were held constant. Measurements for the final height of each plant are provided.

1. Identify the independent and dependent variables in this experiment.

2. Create a data table of the information from the student's experiment.

3. Copy the grid below into your notebook and use the information from the student's experiment to construct a line graph.

a. Use an *x* to plot the final height of each plant at the end of the experiment.

b. Connect the *x*'s with a solid line.

4. Based on your graph or the information provided, determine the expected height of this same type of plant if it were grown in 1 liter of water and 2.5 grams of salt were added.

5. State *one* conclusion, based on the information provided, about the growth of this type of saltwater plant in water containing 0 to 20 grams of salt per liter.

6. In this experiment, the student used plants of the same height and species in equal amounts of water. Identify *one* other condition that the student needed to keep constant.

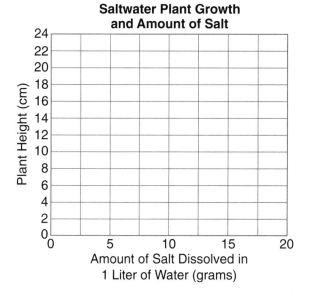

Saltwater Plant Growth and Amount of Salt

1.5 How Do Models Help Us Understand the Natural World?

Objectives

State how models are used in science.

Identify examples of various types of models.

Identify advantages and disadvantages of using models.

Terms

model: a representation of an object or system

scale model: a model in which the relative sizes and positions of all parts are maintained

physical model: a model that you can see or touch

computer model: a model generated by a computer program

simulate: to imitate the functions of a system or process

theory: a detailed explanation of the natural world based on the best information available

Using Models

Sometimes the information that a scientist needs to represent is not numerical in nature. In such cases, a picture really is worth a thousand words. For example, a scientist may draw a diagram that helps us understand the flow of blood through an animal's circulatory system, as in Figure 1.5-1. This diagram does not attempt to show the actual shape of the heart, nor the shapes of the blood vessels. It is a **model** of part of the circulatory system. A model is a representation of an object or system. A model is used to simplify or explain the natural world. Most models are not complete. They represent the important

ideas but leave out the details. Other models allow scientists a way of "playing" without risk. They are a way of doing things by trial and error. Using models enables scientists to make mistakes safely and cheaply. Imagine what the space program would have been like without building models!

What does an atom look like? An atom is a tiny particle that is the building block of matter. Atoms are too small to be seen. Yet you have probably seen the diagram illustrated in Figure 1.5-2. This diagram is an early model of an atom. The model shows us that an atom has two main parts: the nucleus, in the center, and the electrons, moving around it. It does not represent the relative sizes or shapes of these parts of the

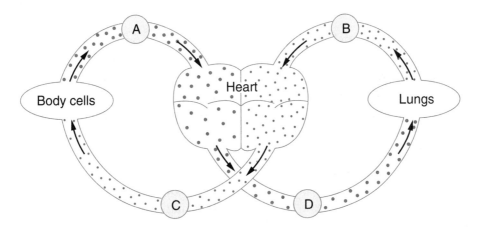

Figure 1.5-1. A model of the human circulatory system.

atom. The model helps us to visualize something that we cannot actually see. Using this model, we can better understand how atoms behave.

Limitations of Models

A model is a representation of reality, but it is not identical to the object or process it represents. There are limitations in the correct use of models. The model of the atom in Figure 1.5-2 is a very old one. Chemists no

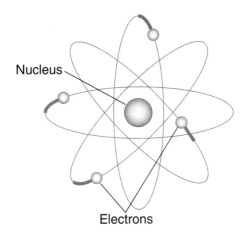

Figure 1.5-2. An early model of the atom.

longer picture atoms as having "rings" of electrons. Even the most modern models have limitations. The nucleus in an atom is so small compared to the rest of the atom that it would be impossible to construct a scale model. A **scale model** is a model in which the relative sizes and positions of all parts of the model are maintained.

Models must be changed over time as new data become available. The earliest model of the solar system had Earth at its center. After the invention of the telescope, more data became available, and it became clear that the sun was the center of the solar system.

Relying too much on models can be dangerous. There could be variables at work that are not considered in the model. On the other hand, *not* having faith in a model can be just as dangerous. Computer models of the levees in New Orleans predicted that they would collapse if a severe hurricane struck that city. Unfortunately, no action was taken, and when hurricane Katrina struck, the city was flooded, just as the models predicted.

Physical Models

You also may have seen a model of the solar system. It shows the sun in the center, with nine planets revolving around it. This model helps us to visualize something very large. The model of the solar system, a globe of Earth, and even a crash dummy, are physical models. A **physical model** is one that you can see and touch. Chemists use molecular models to construct three-dimensional representations of various substances. These models attempt to show how the atoms are arranged in these substances.

A map is another example of a physical model. It can be used to show the factors contributing to the weather. A map can also be used to show the paths of trains. The New York City subway map in Figure 1.5-3 is a physical model you might use every day.

Computer Models

Computer models use special software to **simulate** real situations. Scientists use computer models to predict events such as earthquakes, tsunamis, and the weather. Most storm predictions are based on data that are fed into a computer-modeling program. You even may have heard the weatherperson speak of the "computer projections" when predicting snow accumulations. Forensic scientists use computer models to investigate crime. They attempt to duplicate the conditions of a crime scene with their computers.

Theories

A scientific **theory** is a detailed explanation of the natural world based on the best information available. It is the result of

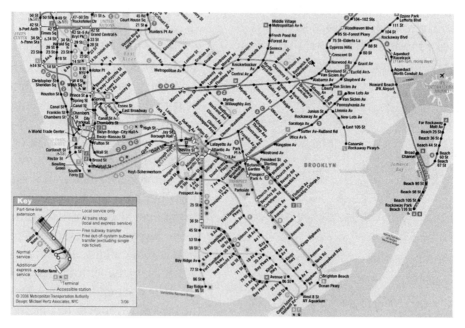

Figure 1.5-3. A map is an example of a physical model.

many experiments and much research. A theory may be considered a third type of model – one that cannot be seen. One important model is the kinetic theory of gases. This model pictures gases as consisting of tiny particles that move randomly within a container. This simple model has been used to explain and predict many of the properties of gases. You will learn about some of these properties in Chapter 3.

As you continue with your science education, you are certain to make use of many models. You will use models of organs, cells, plant parts, Earth, atoms, solar systems, and colliding vehicles. Good models help to make science more interesting and more easily understood by showing you things you couldn't see otherwise.

Questions

1. The items listed below were found in a science classroom.
   - a plastic heart with many of its parts labeled
   - clay formed to look like Earth and other planets
   - a giant plastic plant cell with removable cell parts

   These items are all examples of
   (1) models   (3) variables
   (2) experiments   (4) controls

2. Which statement about physical models is **false**?
   (1) They are exact replicas of the systems they represent.
   (2) They must be changed as new data becomes available.
   (3) They are helpful to scientists.
   (4) They are not always drawn to scale.

3. Which model was most important in determining that the use of airbags in cars would save lives?
   (1) wind tunnels
   (2) crash dummies
   (3) scale models of cars
   (4) maps

Thinking and Analyzing

1. State at least one **advantage** of using models to test new jet planes.

2. State at least one **disadvantage** of relying completely on models.

Review Questions

Term Identification

Each question below shows two terms from Chapter 1. One of the terms is defined.
(1) Choose the term that matches the definition.
(2) Describe how the two terms are different. Following each term is the section (in parenthesis) where the description or definition of that term is found.

1. *Observation (1.1) — Research (1.1)*
   Gathering information, facts, and the opinions of others

2. *Hypothesis (1.2) — Problem (1.2)*
   A possible answer to a problem based on observation or research

3. *Control group (1.3) — Experimental group (1.3)*
   A group in an experiment that receives the independent variable

4. *Independent variable (1.3) — Dependent variable (1.3)*
   A condition in an experiment that is deliberately changed by the scientist

5. *Line graph (1.4) — Bar graph(1.4)*
   A graph that connects numerical data points

6. *Quantitative (1.4) — Scale model (1.5)*
   Involving measured values

7. *Theory (1.5) — Model (1.5)*
   A representation of an object or system

8. *Computer model (1.5) — Physical model (1.5)*
   A model that you can see or touch

Multiple Choice (Part 1)

Choose the response that best completes the sentence or answers the question.

Base your answers to questions 1–4 on the information and figure below and on your knowledge of science.

A gardener performs an experiment growing three different types of plants in equal amounts of soil. Each plant is 10 centimeters tall at the beginning of the experiment. The three plants are given 4 milliliters of water every day for 20 days. The results of the experiment are shown on the graph.

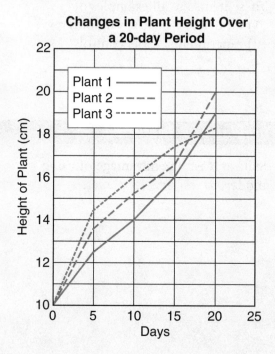

Changes in Plant Height Over a 20-day Period

1. Which plant was the tallest at the end of the 20-day period?
   (1) plant 1
   (2) plant 2
   (3) plant 3
   (4) all plants were the same height

2. Other than at the beginning of the experiment, on what day were plant 1 and plant 3 the same height?
   (1) day 8   (3) day 18
   (2) day 16   (4) day 20

3. Which plant grew at the slowest rate from day 0 to day 5?
   (1) plant 1
   (2) plant 2
   (3) plant 3
   (4) all plants grew at the same rate

4. During which time interval did plant 1 become taller than plant 3?
   (1) days 0-5   (3) days 10-15
   (2) days 5-10   (4) days 15-20

5. The gardener repeated the experiment using identical conditions. Plant 1 grew the fastest. Suggest **one** change to the experimental design that would help the gardener decide which plant really grows fastest.
   (1) Use different amounts of water for each plant.
   (2) Use different amounts of light for each plant.
   (3) Use different amounts of soil for each plant.
   (4) Use many more plants of each type and average the results.

6. Which statement is **always** true about a good model?
   (1) It uses a computer.
   (2) It helps us understand the natural world.
   (3) It is made to scale.
   (4) It can be held or touched.

7. Which of the following is **not** a major field in science?
   (1) Earth science   (3) life science
   (2) physical science (4) pseudo science

8. "What is the effect of temperature on the rate of plant growth?" is best described as a
   (1) problem   (3) variable
   (2) hypothesis   (4) conclusion

9. Which type of graph is used to show the parts of a whole?
   (1) line graph   (3) pie chart
   (2) bar graph   (4) table

10. Using the library and Internet to find reports made by other scientists is called
    (1) research   (3) experimentation
    (2) observation   (4) hypothesizing

11. Which type of graph is most useful to illustrate the following hypothesis?
    "As the temperature of the air increases, the volume of the balloon increases."
    (1) line graph   (3) pie chart
    (2) bar graph   (4) table

12. A scientist
    (1) tests materials for purity
    (2) formulates theories
    (3) does research in the library
    (4) does all of the above

13. Which of the following is correctly paired with its field of science?
    (1) Earth science — botany
    (2) life science — optics
    (3) physical science — chemistry
    (4) Earth science — physics

14. What is the correct order of the scientific method?
    (1) observe→hypothesize→experiment→conclude
    (2) conclude→experiment→hypothesize→observe

(3) hypothesize→conclude→observe→ experiment

(4) experiment→observe→conclude→ hypothesize

15. Which of the following bar graphs represents the same set of data as the pie chart below?

(1) Chart 1   (3) Chart 3

(2) Chart 2   (4) Chart 4

Winners of the World Series from 1996-2005

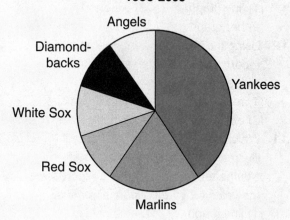

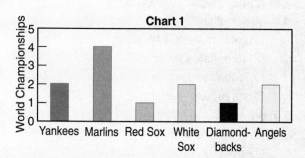

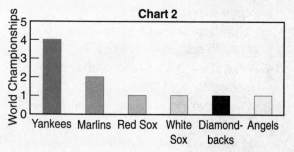

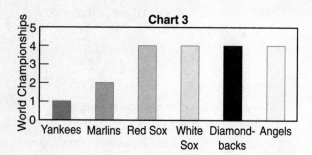

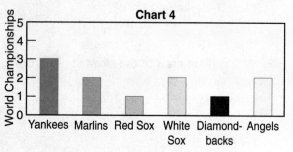

Thinking and Analyzing (Part 2)

Base your answers to questions 1–3 on the information below and the table on page 31 and on your knowledge of science.

A farmer grows and sells flowering plants. The best-selling plants are the ones with the most blossoms. The farmer reads an advertisement for a plant food saying that it will make plants grow faster and taller. The farmer predicts that taller plants will have more blossoms, and performs the following experiment to test this hypothesis.

Two groups of 10 plants each are grown in identical pots filled with equal amounts of identical soil. The amount of sunlight, the room temperature, and the amount of water are held constant for both groups. Group *A* is given plant food at regular intervals according to the instructions on the package. Group *B* is **not** given plant food.

The farmer observes the plants after 15 weeks of growth. The results are recorded on the top of the next page.

Group	Received Plant Food	Average Height (cm)	Average Number of Blossoms
A	yes	35	18.1
B	no	28	18.2

1. State the farmer's original hypothesis.

2. Based on the results of this experiment, is the farmer's original hypothesis correct? Explain your answer.

3. Explain why the amount of sunlight, the room temperature, and the amount of water were held constant for both groups.

Base your answers to questions 4 and 5 on the graph below. The graph shows the results of an experiment that tested the effect of time on the growth of a plant. A student measured the height of several plants for ten weeks and determined the average height. The results are shown in the graph below.

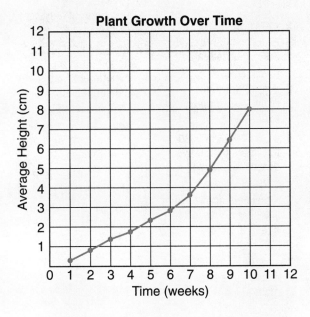

Plant Growth Over Time

4. Identify one variable that should be held constant during this experiment.

5. According to the graph, what will the average height of the plants be at week 11 if growth continues at the same rate as in weeks 8 through 10?

Base your answers to questions 6 through 7 on the information below and on your knowledge of science.

A student determined that shaking a container of sand caused the temperature of the sand to rise. The student then performed a new experiment to see if shaking a container of pebbles causes the temperature of the pebbles to rise.

6. Write a hypothesis for the new experiment.

7. Identify the dependent (responding) variable in the new experiment.

Base your answers to questions 8–10 on the graph below, which shows average yearly temperatures for Earth from 1900 to 1990.

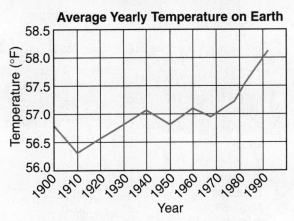

Average Yearly Temperature on Earth

8. Describe what happened to the average yearly temperature on Earth from 1970 to 1990.

9. What was the *lowest* average yearly temperature in the period of time shown?

10. In which 10-year time interval did the average yearly temperature show both an increase and a decrease?

    (1) 1915–1925   (3) 1945–1955
    (2) 1925–1935   (4) 1975–1985

Chapter Puzzle (*Hint:* The words in the puzzle are terms used in the chapter.)

Across

1 involving measured values

5 a model that you can see or touch is called a ____ model

9 the dependent variable in a line graph is always represented by the *y*-____

11 a representation of an object or a system

14 an organized procedure to test a hypothesis

17 visual representations of data, using bars or lines

19 a ____ graph is constructed by connecting data points produced by dependent and independent variables

20 information

21 a detailed explanation of the natural world, based on the best information available

22 the ____ variable is the condition in an experiment that changes due to the effects of the independent variable

23 an organized, step-by-step approach to problem solving is called the scientific ____

24 a system of scientific investigation

Down

2 for clarity, experimental results are organized in a data ____

3 a condition that changes during an experiment

4 a ____ graph is used to show quantitative differences between groups

6 a possible answer to a problem, based on observation or research

7 a model in which relative sizes are maintained is called a ____ model

8 a standard to which the experimental group is compared

10 to imitate the functions of a system or process

12 the ____variable is a condition in an experiment that is deliberately changed by the scientist

13 anything we perceive through one or more of our senses

15 a ____ chart is used to show parts of a whole

16 gathering information, facts, and opinions of others

18 the question to be investigated

20 a model used to test auto safety is called a crash ____

Unit 1

Simple and Complex Machines

Part I—Essential Questions

This unit focuses on two essential questions:

1. How does energy play a role in our lives?
2. How do machines impact our lives?

Energy and the ability to transfer energy play major roles in our lives. How did people travel 150 years ago? They traveled on horseback using "horse energy." They fed a horse and the horse converted the chemical energy in food to mechanical energy. This mechanical energy carried them to their destination. Today we use automobiles, and automobiles convert the chemical energy in gasoline to mechanical energy. Chemical energy that is stored in food and gasoline is converted into mechanical energy of motion that is used to transport us.

Mechanical energy is necessary for just about everything we do. The simple motion of lifting someone on a seesaw, or the complex motion involved in flying an airplane requires a force and work. Machines make work easier by changing 1) the amount of force, 2) the direction of a force required to do work, or 3) the distance over which the force is applied. We use simple machines to transfer mechanical energy with a single motion. A seesaw, a ramp, and a fork are examples of simple machines. Complex machines are made of two or more simple machines. A bicycle, a car, and a wheelbarrow are examples of complex machines.

New, more complex machines will be developed in the future to make our lives even easier. Do you think some day there will be a machine that will get you to your destination faster than a car does today? Perhaps, in the future, there will be a transporter like the one on Star Trek to help us travel from place to place.

Part II—Chapter Overview

In Chapter 2 we discuss the two states of energy—potential and kinetic—describe mechanical energy, and the two types of machines—simple and complex.

Simple machines transfer mechanical energy with a single motion. They make our work easier. There are six types of simple machines: lever, pulley, wheel and axle, inclined plane, wedge, and screw.

Complex machines are made of two or more simple machines. Complex machines transfer other types of energy into mechanical energy. A bicycle is a complex machine that transfers mechanical energy into mechanical energy. A car is a complex machine that transfers chemical energy into mechanical energy. Friction is a force that opposes the motion of machines.

The law of the conservation of energy states that energy can neither be created nor destroyed. However, energy can be transformed from one form of energy into one or more other forms of energy.

Chapter 2

Simple and Complex Machines

Contents

Mechanical energy is transferred from the running water to the turning wheel.

What Is This Chapter About?

Simple and complex machines are important parts of our lives. They help us do work faster and with less effort. Every time we pry a nail from a board, lift someone on a seesaw, ride a bicycle or drive a car, we are using simple and complex machines to transfer mechanical energy.

In this chapter you will learn:

1. Potential and kinetic energy are the two states of energy.

2. There are six simple machines: the lever, pulley, wheel and axle, inclined plane, wedge, and screw. Complex machines contain two or more simple machines.

3. Mechanical energy is the energy of a moving object.

4. Friction is a force that resists motion.

5. The Law of Conservation of Energy states that energy can be neither created nor destroyed.

Career Planning: Engineer

An engineer (en-JA-near) is a trained person who uses scientific knowledge to design, construct, and operate structures and equipment. Some of the many types of engineers are: *aeronautical (air-a-NAW-ti-kal) engineer*—an engineer involved in the design and construction of aircraft; *automotive engineer*—an engineer involved in the design and construction of automobiles; and a *rocket engineer* or *rocket scientist*—an engineer who builds and tests rockets.

Internet Sites:

http://www.edheads.org Edheads—Activities. This Internet site contains a simple machine glossary and some activities that can be done in the classroom or at home.

http://www.mikids.com MIKIDS. This Internet site contains basic information about simple machines. It also has a printable activity sheet, online connections to other simple machine activities, and a quiz.

http://www.beaconlearningcenter.com/WebLessons /MoveOurPrincipal/default.htm. How Can We Move Our Principal? This Internet site provides an interactive activity that teaches students about the six types of simple machines.

2.1 What Are Potential and Kinetic Energy?

Objectives
Define energy.
Describe potential and kinetic energy.

Terms
energy: the ability to do work

work: what is done when a force moves an object a specific distance

potential (puh-TEN-shuhl) **energy:** the stored energy an object has because of its position or condition

kinetic (kih-NEH-tik) **energy:** the energy an object has because it is moving

Energy

Energy is the ability to do work. **Work** is done when a force moves an object over a distance. A flowing river has the ability to move a boat. A car moving down the street can carry people from one place to another. Therefore, the river and the car possess some form of energy.

Potential and Kinetic Energy

There are two basic states of energy: *potential* and *kinetic.*

Potential energy (PE) is stored energy that can be called upon for use at a later time. An object has potential energy because of its relative position or because of its chemical composition. A rock on top of a cliff has potential energy because of its position above ground level. At some future time it can move down the cliff to a lower level. A lump of coal contains potential

energy because it can be burned to produce energy. (See Figure 2.1-1.)

The weight of an object and its height above the surface are used to determine the amount of potential energy an object contains. Potential energy equals weight (newtons) times height (meters):

Potential energy = weight × height

Sample Problem: Which object has the greater potential energy?

Block A – weighs 40 newtons and is 5 meters above the ground.

Block B – weighs 50 newtons and is 6 meters above the ground.

Potential Energy of Block A = 40 N × 5 m or 200 N-m.

Potential Energy of Block B = 50 N × 6 m or 300 N-m.

Block B has the greater potential energy.

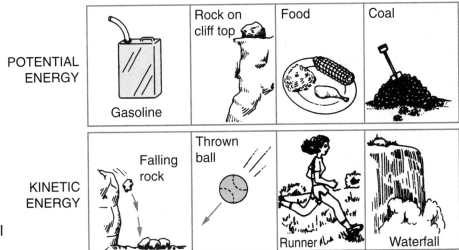

Figure 2.1-1. Some examples of potential and kinetic energy.

Kinetic energy (KE) is energy that an object has because it is moving. (See Figure 2.1-1.) A rock falling off a cliff has kinetic energy. The heat given off by a burning lump of coal is also a form of kinetic energy. The amount of kinetic energy an object contains depends on two things: 1) the weight of the object and 2) the speed of the object. The greater an object weighs, the more kinetic energy it has. The faster an object moves, the more kinetic energy it has. *Stop and Think:* A baseball and a bowling ball are moving at 10 meters/second. Which has the greater kinetic energy? *Answer:* The bowling ball has the greater kinetic energy. Although both balls are moving at the same speed, the bowling ball has the greater weight.

Converting Potential and Kinetic Energy

Potential energy is changed into kinetic energy when an object begins moving. Water held back by a dam has potential energy but no kinetic energy. Releasing the water and letting it flow changes its potential energy into kinetic energy.

Kinetic energy may also be changed into potential energy. (See Figure 2.1-2.) When a ball is thrown straight up into the air, its kinetic energy of motion changes into potential energy as the ball rises higher above the ground. At the highest point of

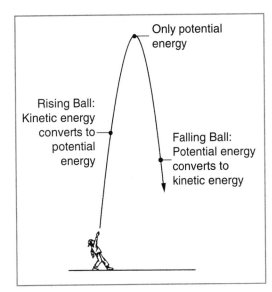

Figure 2.1-2. A ball thrown up in the air demonstrates the conversion of kinetic energy to potential energy, and back again from potential energy to kinetic energy.

Table 2.1-1 Different Forms of Energy

Mechanical energy	is the energy of a moving object. Sound is one type of mechanical energy. It is kinetic energy (KE).
Chemical energy	is energy stored in the bonds of atoms and molecules. It is potential energy (PE).
Nuclear energy	is the energy stored within the nucleus (center) of an atom. It is PE.
Heat energy	is the energy that is associated with vibrating molecules. It is KE.
Electrical energy	is produced by the flow of electrons through a substance. It is KE.
Radiant energy	is a form of energy that moves in waves and can travel through space. Light is a type of radiant energy that is visible to us. It is KE.

the ball's flight, the ball is motionless and has only potential energy. As the ball falls back to the ground, the potential energy changes back into kinetic energy.

Forms of Energy

There are six forms of energy. Remember, each form of energy has the ability to do work, or move an object. Table 2.1-1 lists the different forms of energy.

Stop and Think: Can you give an example of each form of energy that is commonly found around the house? Which form of energy is *not* commonly found around the house? *Answer:* Mechanical energy – turning hands on a clock. Chemical energy – a battery. Nuclear energy – *not* commonly found around the house. Heat energy – from the stove. Electrical energy – working television set. Radiant energy – turned-on lamp light.

Activity

1. List five objects at home or school that contain kinetic energy, and five objects that contain potential energy.

2. Using a ball and a ramp as shown below, develop a demonstration that will show 1) the ball with maximum potential energy, 2) converting potential energy into kinetic energy, and 3) the ball with maximum kinetic energy.

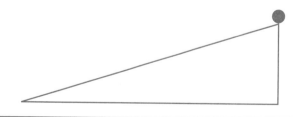

SKILL EXERCISE—*Interpreting a Diagram*

It may not be obvious, but a roller coaster does not have an engine. The car is pulled to the top of the first hill by a motor and pulley system. At the top of the first hill, it contains sufficient potential energy to complete the ride.

Throughout the ride, different wheels on the car keep the car moving without falling off the track. Three sets of wheels guide the car on the track. Running wheels keep the car moving forward. Friction wheels control sideway movement. The third set of wheels keep the car on the track even when it is upside down.

The diagram below shows a profile view of a roller coaster. The letters A–F indicate six positions along the profile.

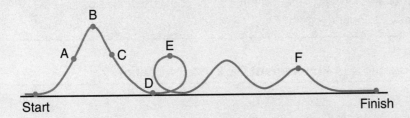

Based on this profile, answer the following questions:

1. Where is the potential energy the greatest?
2. Where is the kinetic energy changing into potential energy?
3. Where is the potential energy changing into kinetic energy?
4. Where is the kinetic energy the greatest?

This Internet site, *http://www.fearofphysics.com* (go to Visual Physics and then Roller Coasters) discusses potential and kinetic energy during a roller coaster ride. The site is interactive and allows you to select your own roller coaster arrangement and try it out.

Interesting Facts About Roller Coasters

The first roller coasters were probably built in Russia in the 1700s. These early roller coasters consisted of a steep slide made of ice. By the 1800s, they also appeared in France. Soon after, a warm-weather version was made using wood and rollers. In the late 1800s, indoor roller coasters appeared in the United States. The ride consisted of a sled and rollers.

More than 2000 wood-constructed roller coasters were built in the United States from 1900 – 1930. They were very popular. However, during the depression and World War II many people stopped going to amusement parks. Steel roller coasters became popular in the 1960s. However, wood roller coasters once again became popular in the 1970s. A variety of exciting roller-coaster features also started to gain popularity at this time. Today, we have inverted (upside-down) roller coasters, multiple-looping, roller coasters, stand-up roller coasters, and suspended roller coasters. Who knows what's coming next?

The chart below lists the top three roller coasters ranked by speed, height, drop, and length.

Rank		Roller Coaster	Amusement Park	Location
	Speed			
1st	128 mph	Kingda Ka	Six Flags Great Adventure	Jackson, New Jersey, USA
2nd	120 mph	Top Thrill Dragster	Cedar Point	Sandusky, Ohio, USA
3rd	106.9 mph	Dodonpa	Fuji-Q Highland	Fujiyoshida, Shizuoka, Japan
	Height			
1st	456'	Kingda Ka	Six Flags Great Adventure	Jackson, New Jersey, USA
2nd	420'	Top Thrill Dragster	Cedar Point	Sandusky, Ohio, USA
3rd	415'	Superman The Escape	Six Flags Magic Mountain	Valencia, California, USA
	Drop			
1st	418'	Kingda Ka	Six Flags Great Adventure	Jackson, New Jersey, USA
2nd	400'	Top Thrill Dragster	Cedar Point USA	Sandusky, Ohio, USA
3rd	328' 1"	Superman The Escape	Six Flags Magic Mountain	Valencia, California, USA
	Length			
1st	7677' 2"	Daidarasaurus	Expoland	Suita, Osaka, Japan
2nd	7442'	Ultimate	Lightwater Valley	Ripon, N. Yorkshire, UK
3rd	7359'	Beast	Paramount's Kings Island	Kings Mills, Ohio, USA

Source: *http://www.rcdb.com/rhr.htm*

Questions

1. The diagram below shows a boulder rolling down a hill into a valley and then up the opposite side.

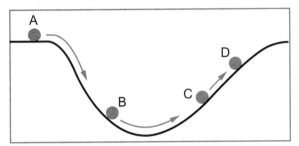

At which position does the boulder have the greatest kinetic energy?
(1) A   (3) C
(2) B   (4) D

2. A child sits at the top of a playground slide and then slides down. This is an example of changing
(1) work to kinetic energy
(2) kinetic energy to work
(3) kinetic energy to potential energy
(4) potential energy to kinetic energy

3. The diagram below shows two balls on platforms. What can be said about the potential energy of the balls?

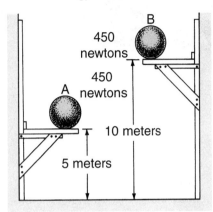

(1) Ball A has a greater potential energy.
(2) Ball B has a greater potential energy.
(3) Both balls have the same potential energy.
(4) Neither ball has potential energy.

Thinking and Analyzing

A pendulum is a suspended object that is allowed to swing back and forth. (See the figure to the right.) This allows the object (in this case a ball) to swing to-and-fro when it is pushed. The figure shows five positions of a pendulum as it swings from left to right.

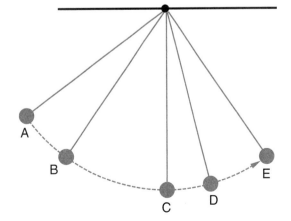

a. At which position is the potential energy the greatest?

b. At which position is the potential energy the least?

c. At which position is the kinetic energy the greatest?

What Are Simple Machines?

Objectives

Describe how machines help us.

Describe the different types of simple machines.

Terms

machine: a device that transfers mechanical energy from one object to another

resistance (ri-ZIS-tans): the force a machine has to overcome

effort: the force applied to a machine

fulcrum (FUHL-krum): the point around which a lever turns

How Machines Help Us

A **machine** is a device that transfers mechanical energy from one object to another object. Machines make work easier. They do this by changing

1) the amount of force,
2) the direction of a force, or
3) the distance over which a force is applied.

The force a machine has to overcome is called **resistance**, and the force applied to a machine is called **effort**. A machine can reduce the amount of effort needed to overcome a large resistance.

A wrench increases the force (effort) applied to it when removing a tight bolt (resistance). The tight bolt is the resistance force. However, the distance the effort is applied is much greater than the distance the bolt turns. (See Figure 2.2-1.)

A *simple machine* is a device that uses only one motion to transfer mechanical

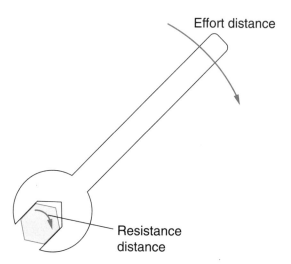

Figure 2.2-1. A small effort at the end of a wrench is capable of overcoming a great resistance of a tight bolt. The distance the effort is applied is greater than the distance the bolt turns.

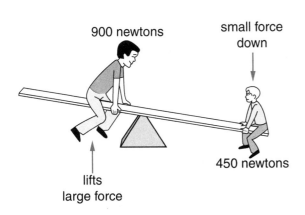

Figure 2.2-2. A seesaw is a simple machine. A small effort force down can lift a larger resistance force.

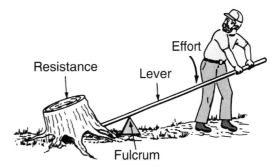

Figure 2.2-3. A lever multiplies effort, making it easier to uproot a tree stump.

energy. A seesaw is a simple machine. A small downward force on one side of the seesaw can lift a large force on the other side. (See Figure 2.2-2.) Where is the effort and where is the resistance in the seesaw in the figure?

Types of Simple Machines

Simple machines help us do work faster and with less effort. There are six types of simple machines: the lever, pulley, wheel and axle, inclined plane, wedge, and screw.

A *lever* is a rigid bar that turns around a point called a **fulcrum.** (See Figure 2.2-3.) Levers make work easier by increasing the applied force, or effort. Scissors and crowbars are examples of levers.

A *pulley* is a simple machine consisting of a rope and one or more wheels. The rope fits into a groove around the wheel(s). The resistance is a weight attached to one end of the rope or to one of the wheels. The effort is applied to the other end of the rope. A single-wheel pulley is used to change the direction of a force. It can be used to raise a flag up a flagpole. Multiple-wheel pulleys can increase the effort. A heavy car engine can be lifted using a multiple-pulley system. Figure 2.2-4 shows several types of pulleys.

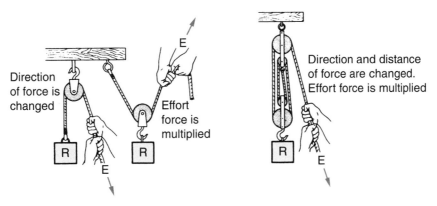

Figure 2.2-4. Three types of pulleys. *R* stands for resistance, and *E* means effort.

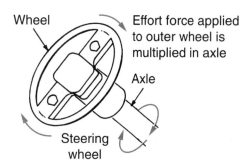

Figure 2.2-5. A steering wheel is an example of a wheel and axle.

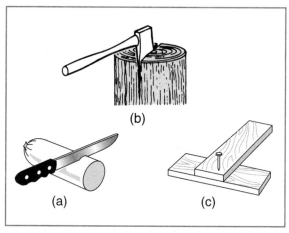

Figure 2.2-7. Three examples of wedges: (a) knife, (b) ax, and (c) wood nail.

A *wheel and axle* is a large wheel with a smaller wheel, or axle, in its center. The wheel is attached to the axle so they turn together. When the outer wheel is turned, so is the axle. (See Figure 2.2-5.) The outer wheel is turned a greater distance, but uses less force. Bicycle handlebars, steering wheels in cars, and doorknobs are examples of a wheel and axle.

An *inclined plane* is a flat surface with one end higher than the other. A wheelchair ramp and a truck ramp are inclined planes. A staircase is another example of an inclined plane. Figure 2.2-6 shows how an inclined plane makes work easier by decreasing the force needed, while increasing the distance the force is applied.

The *wedge* is a double-sided inclined plane. The effort is applied by driving the wedge into something. For example, when an ax is driven into a log, the ax splits open the log. Other examples of wedges are knives, nails, and chisels. (See Figure 2.2-7.)

The *screw* is an inclined plane wrapped around a wedge or cylinder. Examples of screws are wood screws, bolts, and car jacks.

Figure 2.2-6. A loading ramp is an inclined plane.

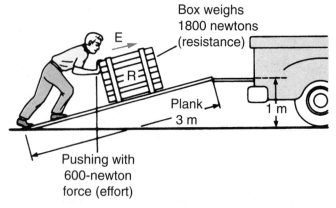

Figure 2.2-8. A wood screw is a spiral inclined plane.

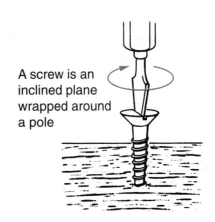

A screw is an inclined plane wrapped around a pole

When using a wood screw, a circular force is applied to the screw's head in order to overcome the resistance of the wood. As the screw turns, it penetrates the wood. (See Figure 2.2-8.)

Activities

1. Construct a chart listing the six types of simple machines and list two examples found around the home or in school.

2. Place a wooden ruler near the edge of a table with about 5 centimeters extending over the edge. Place a small book on the part of the ruler that is on the table. Lift the book by pressing down on the end of the ruler that extends beyond the edge of the table, as shown in the figure. Repeat the procedure with the ruler, extending it 10 centimeters, 15 centimeters, and then 20 centimeters over the edge of the table. (a) Describe the type of simple machine demonstrated in this activity. (b) Where are the effort, fulcrum, and resistance? (c) When is it easiest to lift the book?

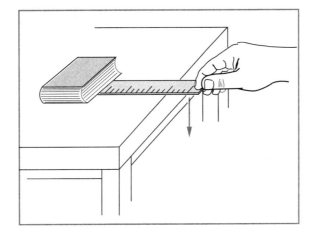

SKILL EXERCISE—*Designing an Observation Procedure*

Many common tools are levers of some kind. For example, scissors, shovels, and salad tongs are levers. Levers can be grouped into three basic classes, depending on where the lever encounters the resistance, where the effort is applied, and where the fulcrum (the point around which the lever turns) is located. Figure 2.2-9 illustrates the three classes of levers.

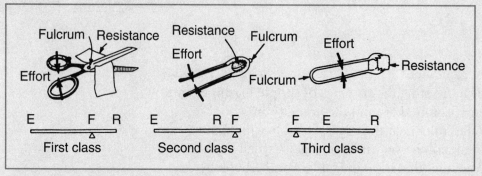

Figure 2.2-9. Three classes of levers.

A first-class lever, such as a pair of scissors, has the effort (E) applied on one end, the resistance (R) on the other end, and the fulcrum (F) in between. A second-class lever, like a nutcracker, has the fulcrum and effort at opposite ends, and the resistance in the middle. A third-class lever, such as a pair of ice tongs, has the resistance and the fulcrum at opposite ends, and the effort applied in the middle.

Figure 2.2-10 shows some examples of levers. Can you determine which lever class each item represents? Drawing a lever diagram for

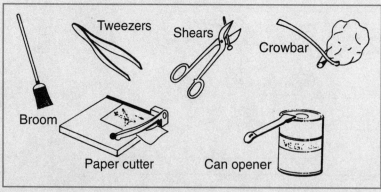

Figure 2.2-10. Examples of levers.

each one will help you decide. First, draw a line to represent the lever. Next, think about how you use the lever, and try to identify the positions of the fulcrum, effort, and resistance. Where does the lever meet resistance? Is effort applied to one of the ends of the lever, or somewhere in between? Where does the lever turn or change direction? Once you have located the fulcrum, resistance, and effort, and labeled them on your lever diagram, you can then classify the lever using the definitions of lever classes given before.

Figure 2.2-11 shows how this would be done for a broom. The lever diagram fits the definition of a third-class lever. Do the other items on your own, and then answer the following questions.

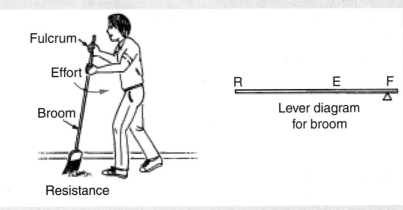

Figure 2.2-11. Using a broom, and its lever diagram.

1. The wheelbarrow is a type of lever. The wheel in front is the location of the
(1) effort   (2) fulcrum
(3) resistance (4) force

2. Which of the following levers is the type shown in the lever diagram?
(1) scissors   (2) paper cutter
(3) tweezers   (4) bottle opener

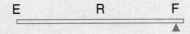

Interesting Facts About Pyramids

Building the Pyramids

Egyptian pyramids were built about 4500 years ago, long before modern machines. They were built as tombs for pharaohs (FER-os), or kings of ancient Egypt. The Great Pyramid near Gizeh is about 150 meters high and has twice the volume of the Empire State Building. Even with a tremendous work force of thousands of laborers, it took many years and creative thinking to cut and move the great number of 2 1/2 ton limestone blocks needed to build the pyramids.

It is believed that much of the work was done using ramps and rounded logs as rollers under the massive stone blocks. The use of a slick, muddy surface under the stones may also have aided the sliding of the blocks by reducing friction. It is believed that levers and ropes were used to lift and place the stones. Although the exact method of building pyramids is not known, we know that to build the pyramids the Egyptians used simple machines like the inclined plane, the wheel, and the lever.

Questions

1. A machine is a device that transfers
   (1) electrical energy from one object to another object
   (2) mechanical energy from one object to another object
   (3) electrical energy and mechanical energy between objects
   (4) heat energy between objects

2. The drawing below shows a person about to lift a book using a ruler and pencil. The drawing illustrates an example of which simple machine?
   (1) gear   (3) balance
   (2) pulley   (4) lever

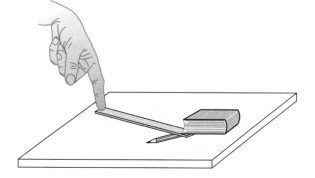

3. A student in a wheelchair is approaching the door to the school by going up a ramp, as shown in the figure below. Which two simple machines are being used to enable the student to reach the door?

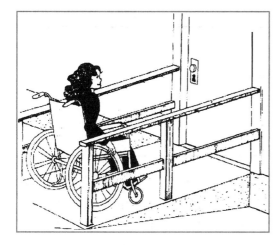

   (1) inclined plane and pulley
   (2) lever and wheel and axle
   (3) pulley and lever
   (4) wheel and axle and inclined plane

4. A seesaw is a simple machine because it transfers
   (1) mechanical energy with one motion
   (2) mechanical energy into heat energy
   (3) mechanical energy with two motions
   (4) heat energy with one motion

5. The pulley on the flagpole in the illustration makes it easier to raise the flag by
   (1) decreasing the amount of work required
   (2) changing the direction the force is applied
   (3) decreasing the force required
   (4) making the flag lighter

Pulley

Thinking and Analyzing

You have a box that weighs 450 newtons and you want to place it on a shelf 1 meter above the floor. The box is on the floor 5 meters away from the shelf. You are not capable of lifting the box. Using simple machines, how might you place the box on the shelf?

What Are Complex Machines?

Objectives
Define a complex machine.

Describe examples of complex machines.

Term
complex machine: a machine made of two or more simple machines

---

Complex Machines

Most machines used in your daily life are **complex machines**. They are made up of two or more simple machines. Some books refer to them as *compound machines*.

We know a simple machine makes work easier. Complex machines can do jobs easier and faster than simple machines. For example, a vacuum cleaner is a complex machine. It takes much less time and effort to remove dirt from the floor and rugs with a vacuum cleaner than with a broom.

Examples of Complex Machines

Scissors are complex machines that contain two simple machines. The two levers are joined together by a screw that acts as a fulcrum. Along each lever there is a wedge, or cutting blade.

A hand can opener is an example of a complex machine containing three simple machines. It contains a wedge, lever, and wheel and axle. The wedge is on the small round blade that cuts into the metal can.

The lever applies the forces to push the wedge into the can. You turn the wheel and axle so the can opener easily cuts the lid around the top of the can.

A bicycle is a complex machine. It is made of at least four different simple machines. (See Figure 2.3-1.) Table 2.3-1 lists four simple machines found on a bicycle.

A car is a highly complex machine that contains many simple machines. A few of the simple machines found in a car are:

- a door, which is a lever
- a fan belt, which is a pulley

Table 2.3-1. Four Simple Machines Found on a Bicycle

Bicycle Part	Simple Machine
Handbrakes	Lever
Handlebars	Wheel and axle
Chain	Pulley
Wheel bolts	Screw

Figure 2.3-1. Can you locate simple machines within a bicycle?

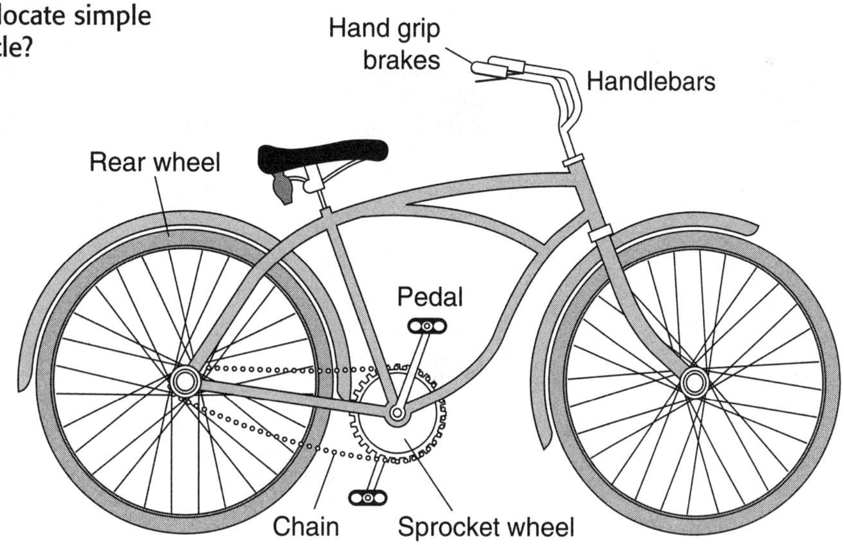

- a manual adjustable seat that uses a lever
- a manual window opener, which is a wheel and axle
- a car jack, which is a screw

Growth of Machines

Simple machines have been used by humans since early times. Primitive humans made tools and weapons from stone. Arrowheads, knifes, and axes were wedges designed for hunting and cutting. Ancient civilizations such as the Egyptians, Greeks, and Romans used inclined planes, the wheel and axle, and pulley systems to create monuments and buildings. During the agricultural revolution, simple machines were improved for plowing the soil and complex machines were used to help pick the crops. New complex machines were invented during the Industrial Revolution to weave fabric and mass produce items in factories.

Today, complex machines have "brains," or computer chips to tell them what to do. Robots are complex machines that are capable of cleaning sewers, laying cable, checking for dangerous explosives, and even doing surgery. It appears that robots will some day be capable of performing many other jobs, too.

Activities

1. Walk around your house and make a list of at least five complex machines. Name two simple machines that each of the complex machines contain.
2. Find a construction site near your home and observe the workers using heavy equipment to dig, move, and lift construction materials. Choose one piece of heavy equipment and see how many simple machines you can identify in it.

SKILL EXERCISE—*Analyzing Complex Machines*

Even though you have learned that complex machines contain two or more simple machines, sometimes it is difficult to identify what simple machines a complex machine contains. In this exercise, you will be shown three complex machines. A description of the purpose of the machine and how it operates is given below. You are to determine what simple machines are contained in each of the complex machines.

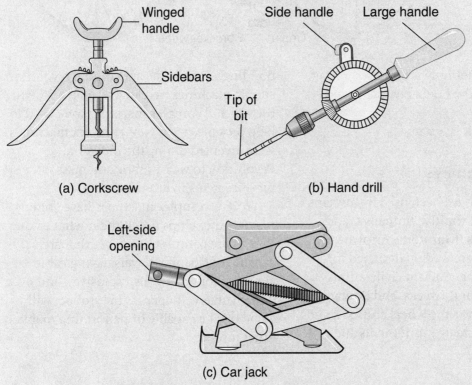

Figure 2.3-2. Three complex machines.

a) Corkscrew: This item is used to remove corks from bottles. Place the corkscrew into the cork at the top of the bottle. Turn the winged handle on top until the corkscrew is fully embedded in the cork. The two long sidebars move up. Push down on the two sidebars and the cork comes out of the bottle. What three simple machines does a corkscrew contain?

b) Hand Drill: It is used to drill a hole in wood. Place the tip of the bit on the wood where the hole will be drilled. Hold the large, long handle with one hand and turn the side handle with the other hand. What two simple machines does a hand drill contain?

c) Car Jack: It is used to lift a car to change a tire. Place the jack under the car. Place a steel rod in the left side opening and turn the rod. The two sides of the jack come together lifting the car on top of the jack. What two simple machines does the car jack contain?

Questions

1. A machine that contains two or more simple machines is called a
   (1) industrial machine
   (2) complex machine
   (3) multiple machine
   (4) perpetual machine

2. Compared to simple machines, complex machines
   (1) make the job more difficult
   (2) make the job easier
   (3) take more time
   (4) take more effort

3. A knife cutting a sandwich in half is an example of a complex machine. What simple machines does a cutting knife demonstrate?
   (1) an inclined plane and pulley
   (2) an inclined plane and screw
   (3) a wedge and lever
   (4) a wedge and screw

Thinking and Analyzing

1. Explain why a manual pencil sharpener is a complex machine.

2. What two simple machines are being used when a person uses a hand truck to move heavy boxes, as shown in the figure to the right?

2.4 How Do Machines Affect Work and Force?

Objectives

Describe the relationship between work and force.

Explain why work input is always greater than work output in a machine.

Terms

newton: a unit of force in the metric system (about 0.22 lb, or 1 lb = about 4.5 newtons)

newton-meter (N-m): a force of one newton acting over a distance of one meter

joule (JOOL): the amount of work done by a force of one newton over a distance of one meter; one joule equals one N-m

efficiency (e-FI-shan-see): the ratio of work output to work input

---

Work

Scientists use the word "work" in a very specific way. They say that *work* is done when a force causes an object to move some distance. The amount of work done depends on the amount of force applied and the distance the object is moved. The relationship among work, force, and distance is given by the formula:

Work = Force $\times$ distance, or W = F $\times$ d

When a force is applied to an object, the force may or may not cause the object to move. If the force does not produce motion, no work is done. A force results in work only if motion is produced. (See Figure 2.4-1.)

You are probably familiar with the unit of force called a pound (lb). The **newton** is a unit of force in the metric system (about 0.22 lb, or 1 lb = about 4.5 newtons). The

unit of work is the **newton-meter (N-m).** One newton-meter of work is done when an object weighing 1 newton is lifted 1 meter

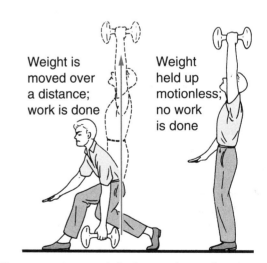

Weight is moved over a distance; work is done

Weight held up motionless; no work is done

Figure 2.4-1. Work is done when a force acts over a distance. When an object is held motionless, no work is being done.

high. One newton-meter is also called a
joule.

   A sample work problem demonstrates
how to determine the amount of work done.
A girl weighing 400 newtons walks up a 3-
meter-high staircase. How much work has
she done?

   W = F × d
   W = 400 newtons × 3 meters
   W = 1200 N-m, or 1200 joules

   Now you try one: How much work is
done when a weight lifter lifts a 1000-
newton barbell from the floor over his head
to a height of 2.2 meters?

Machines and Work

Ideally, a machine's work output should
equal the amount of work put into a
machine. However, machines are never 100
percent efficient. The amount of work put
into a machine is always greater than the
amount of work done by the machine. This
is true because some of the work put into a
machine is changed into heat energy that is
usually wasted. The heat is produced by
friction that is caused by moving parts
rubbing together in the machine.

   Ideally—work input *equals* work output
   Actually—work input is *greater* than
         work output

   An example will make this clearer.
Suppose you use pulleys to lift a 500-
newton box to a height of 2 meters. (See
Figure 2.4-2.) Although 1000 N-m (500 N ×
2 m) of work output was done, you actually
had to perform more than 1000 N-m of
work input. Some of your work input is
wasted, because heat is produced by the

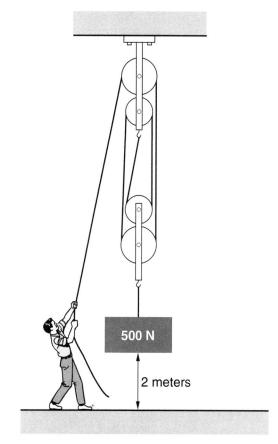

Figure 2.4-2. A pulley system is used to lift a
500-N object two meters high.

friction between the wheel and axle of each
pulley, and between the rope and wheel of
each pulley. So, in reality, the work input is
greater than 1000 N-m.

Efficiency of a Machine

The **efficiency** of a machine is measured by
the percentage of work output compared to
the work input. As a formula, efficiency can
be stated:

$$\text{Efficiency} = \frac{\text{Work Output}}{\text{Work Input}} \times 100$$

If a simple machine has a work input of 600 N-m, and a work output of 300 N-m, the efficiency would be determined by:

$$\text{Efficiency} = \frac{\text{Work Output}}{\text{Work Input}} = \frac{300}{600} \times 100$$
$$= 50 \text{ Percent}$$

If the amount of friction in a simple machine decreases, the work input will decrease and the efficiency would increase. In the previous problem, if the work input decreased to 400 N-m, the efficiency of the simple machine would be 75 percent.

Interesting Facts About Machines

A few centuries ago inventors dreamed about building a perpetual motion machine. A perpetual motion machine is a machine that, once set in motion, would continue its motion forever without further energy input. Spin a wheel on a toy car. It eventually stops spinning, but a perpetual wheel would continue to spin without stopping. Hundreds of perpetual motion machines were designed and built, but not one worked.

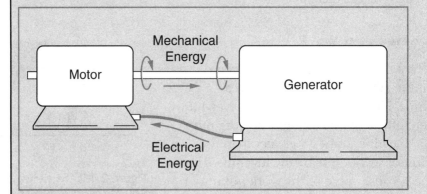

Figure 2.4-3. A design of a perpetual motion machine. The motor uses electricity to produce mechanical energy, and the generator uses mechanical energy to make electricity.

An interesting perpetual motion machine was proposed in the late 1800s. A motor changes electrical energy into mechanical energy, and a generator changes mechanical energy into electrical energy. If you connect a motor and a generator, the output from the motor should turn the generator, and the turning generator should produce electricity to power the motor. (See Figure 2.4-3.) Once again, when this device was built, it failed.

The concept of the perpetual motion machine is attractive. Invent it and you will become rich and famous. Unfortunately, it is an impossible mission. Do you know why it is impossible to build a perpetual motion machine? Your knowledge of the efficiency of machines should help you with the answer.

Questions

1. Work is done only if a force is applied and
   (1) there is motion
   (2) there is no motion
   (3) it is done in a short period of time
   (4) the force is great enough

2. What is a unit of work?
   (1) newtons   (3) meters
   (2) newtons/sec   (4) newton-meters

3. Jessie lifted a box weighing 40 newtons and placed it on a shelf 1 meter high. Sara lifted a second box weighing 50 newtons and placed it on the same shelf. Which statement correctly describes this situation?
   (1) Jessie did more work than Sara.
   (2) Jessie used a greater force than Sara.
   (3) Sara and Jessie did the same amount of work.
   (4) Sara did more work than Jessie.

4. The relationship of work input and work output in a simple machine is
   (1) work input is always equal to work output
   (2) work input is always less than work output
   (3) work input is always greater than work output
   (4) work input is sometimes greater than work output

5. The efficiency of a simple machine is always less than 100% because
   (1) some of the work input is lost to force
   (2) some of the work input is lost to effort
   (3) some of the work input is lost to friction
   (4) some of the work input is lost to resistance

Thinking and Analyzing

Use the formulas on pages 54 and 55 to answer the following questions.

1. A student uses a pulley system to raise a 100-newton weight 10 meters. The student uses 1250 N-m of work to raise the weight.
   (a) What is the work input?
   (b) What is the work output?
   (c) How much work is lost in the process of raising the weight?
   (d) What is the most likely cause of the lost work?
   (e) What is the efficiency of the pulley system?

2. A 6th grade student who weighs 500 N climbed the stairs in his apartment building. The stairs brought him 15 meters above the ground floor.
   (a) How much work did the student do?
   (b) If the student carried schoolbooks that weigh 20 N up the flight of stairs, how much additional work was done?
   (c) How many pounds did the student weigh?

2.5

What Is Mechanical Energy?

Objectives

Define mechanical energy.

Explain how different forms of energy can change into mechanical energy.

Describe how simple machines use mechanical energy.

Term

mechanical (muh-KA-nih-kuhl) **energy:** the energy of a moving object; it is kinetic energy (KE)

Mechanical Energy

Mechanical energy is the energy of a moving object. It is the form of energy we are most familiar with in our daily lives. When we ride a bicycle, we use mechanical energy to turn the pedals. The turning pedals transfer mechanical energy to the chain, which transfers mechanical energy to the back wheel, and the bicycle moves forward.

A turning electric motor produces mechanical energy to operate power tools and home appliances. The rotary blade of a power saw turns and cuts wood. Spinning blades in a kitchen blender are used to mix food. A washing machine moves clothes back and forth in an effort to remove the dirt. *Stop and Think:* What other examples of mechanical energy can you identify from school or in your home? *Answer:* Some examples of mechanical energy from school are the ringing bells, a thrown ball on the playground, and a pencil falling to the floor. Some examples at home are a rotating fan, hands turning on a clock, and the wind blowing the curtain.

Producing Mechanical Energy

To produce mechanical energy it is necessary to obtain energy from another energy source. The following are examples of how different forms of energy are changed into mechanical energy:

Electrical Energy to Mechanical Energy — A ceiling fan changes electrical energy into a turning fan. The motor in a ceiling fan uses electrical energy to produce mechanical energy in the form of turning blades.

Chemical Energy to Mechanical Energy — A car changes the chemical energy in gasoline to the movement of a car. An automobile is able to speed down the highway because its engine changes the chemical energy in gasoline to mechanical energy.

Heat Energy to Mechanical Energy — A steam engine changes heat energy into mechanical energy that causes the turning motion in electrical generators.

Mechanical Energy to Mechanical Energy — The mechanical energy of the wind is transferred to the mechanical energy of a spinning pinwheel.

Light Energy to Mechanical Energy — A radiometer can change light energy into mechanical energy that causes the turning motion of paddles. (See Figure 2.5-1.)

Figure 2.5-1. Light energy causes the paddles to turn inside the bulb (radiometer).

Activity

Make a copy of the chart below on a sheet of paper. In the left column of the chart, list 10 items that demonstrate mechanical energy. In the right column, state the form of energy that was used to produce the mechanical energy. See the example. A battery-operated bunny changes the chemical energy in the battery to mechanical energy of a hopping bunny rabbit.

Item Demonstrating Mechanical Energy	Form of Energy Used to Produce Mechanical Energy
Example: Battery-Operated Bunny Rabbit	Chemical Energy
1.	
2.	
3.	
4.	
5.	
6.	
7.	
8	
9.	
10.	

Mechanical Energy and Simple Machines

We already learned that a simple machine is a device that uses only a single motion to transfer mechanical energy. In other words, mechanical energy goes into a simple machine and mechanical energy comes out of a simple machine. An example of how mechanical energy is transferred in each type of simple machine is illustrated in Figure 2.5-2.

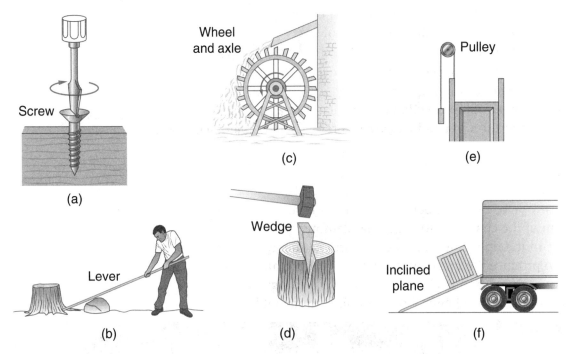

Figure 2.5-2. Examples of how mechanical energy is transferred in simple machines.

(a) Screw: Turning a screw causes it to pull into the wood.
(b) Lever: The downward motion of a lever lifts a tree stump.
(c) Wheel and axle: Running water turns a water wheel.
(d) Wedge: The pounding of a wedge into a log causes the log to split.
(e) Pulley: Applying a downward force on the rope of a pulley raises an object.
(f) Inclined plane: Pushing a crate up an inclined plane causes the crate to move.

Questions

1. Which item demonstrates mechanical energy?
   (1) a lightbulb   (3) a falling baseball
   (2) a mirror   (4) gasoline

2. Mechanical energy in an object can be obtained from another energy source. The mechanical energy in a clock is commonly obtained from
   (1) electrical energy
   (2) sound energy
   (3) light energy
   (4) heat energy

3. An example of chemical energy changing into mechanical energy is
   (1) a ceiling fan
   (2) a boat on a river
   (3) a battery-operated toy
   (4) a burning log

4. A claw hammer has a forked end that is used for removing nails. When you use a claw hammer to pry a nail out of a piece of wood, the mechanical energy that goes into the process is
   (1) the nail coming out of the wood
   (2) the nail
   (3) the hammer
   (4) the motion of the hammer

5. Baseball player A throws a ball to player B. Player B catches the ball. Player A puts mechanical energy into the ball with his arm. The moving ball contains mechanical energy. Eventually, the mechanical energy is transferred
   (1) into the air
   (2) into the ground
   (3) into player B's glove
   (4) into player B's arm

Thinking and Analyzing

1. There are many examples of how an object with mechanical energy causes another object to move. Give an example of how each of the following objects uses mechanical energy to move another object.
   (a) hammer
   (b) ceiling fan
   (c) shovel
   (d) car jack
   (e) bicycle handlebars

2. For each of the three diagrams below, state what simple machine is demonstrated. Also, describe the mechanical energy that goes into the simple machine, and the mechanical energy that comes out.

(a)

(b)

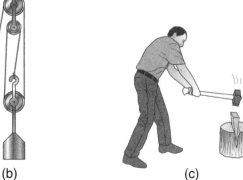

(c)

2.6 How Does Friction Affect Machines?

Objectives

Describe how friction affects the motion of an object.

Describe how friction can be reduced.

Explain the effect of friction in simple machines.

Term

friction (FRICK-shun): a force that resists the motion of an object when one surface moves over another surface

---

Friction Is a Force

Friction is a force that resists motion. It is created when one surface slides across another surface. Friction must be overcome to move an object at rest and to keep an object moving. For example, the force needed to start sliding a book on a tabletop must be great enough to overcome frictional resistance. Once moving, friction between the book and the tabletop slows the book's motion. Eventually, the frictional force between the moving book and the tabletop causes the book to stop.

The rougher the surface and the heavier the object, the greater the frictional force. Friction makes it more difficult to move an object. Pushing a book across a smooth tabletop is easier than pushing a book across a rough tabletop. Pushing a heavier book requires a greater force than pushing a lighter book. (See Figure 2.6-1.)

Figure 2.6-1. Book A weighs 4 newtons and Book B weighs 8 newtons. If both books are slid across the table with the same force, Book A will travel farther.

The Effect of Friction on Motion and Work

We commonly think of friction as undesirable, especially when it interferes with motion and work. When we try to move an object, the force of friction resists the motion of the object. Sliding a heavy box across the floor may take a great amount of effort. If the friction between the bottom of the box and the floor is reduced, less effort is necessary to move the box.

But friction can also be a very desirable force. Car tires have irregular surfaces (called treads) that increase the friction with the road. Friction causes *traction* between a car's tires and the road, and allows a car to start moving, make turns, and come to a stop.

An icy surface has very little friction. Think about how a car starts and stops on ice. *Stop and Think:* What happens if you try to walk on a frictionless surface, like ice? *Answer:* If you try to walk on a frictionless surface like ice, your feet slip and slide, causing you to fall. You need friction so you can walk without slipping and sliding.

Decreasing Friction

Sometimes we try to decrease friction to make work easier. Some ways to decrease friction include using: (1) lubrication; (2) ball bearings; (3) polished surfaces; and (4) rollers. (See Figure 2.6-2.)

Table 2.6-1 lists examples of ways to decrease friction.

Table 2.6-1 Decreasing Friction

Method	Examples	Where Used
Lubrication	Oil, grease, graphite	Wheel and axle, doorknob
Ball bearings	Small metal balls	Roller blades, car wheels
Polished surface	Smooth surface	Playground slide, sled
Rollers	Wheels	Hand truck, furniture rollers

Friction in Simple Machines

Friction decreases the efficiency of a simple machine. Decreasing friction at the friction point of a simple machine (see Table 2.6-2) improves its efficiency. This means less force is needed to produce the desired work. For example, you need less force—it is easier— to slide a box up an inclined plane once you have made it smoother.

Table 2.6-2. Friction Points in Simple Machines

Simple Machine	Friction Point
Lever	at the fulcrum
Pulley	axle of pulley and between rope and pulley
Wheel and axle	on the axle
Inclined plane	along the surface of the inclined plane
Wedge	between surface of wedge and object being split
Screw	between thread of screw and substance screw enters

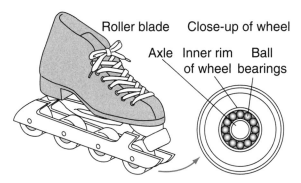

Figure 2.6-2. Ball bearings in the wheel of a roller blade reduce friction as the wheel turns.

SKILL EXERCISE—*Investigating Friction*

*T*o do this activity you need 1) a 2–3-mm-thick rubber band; 2) a block of wood with one flat side (about 15 cm x 10 cm x 5 cm); 3) a hook that screws into wood; 4) a metric ruler; and 5) five different flat surfaces (for example: sandpaper, carpet, glass, floor tile, and concrete). Cut the rubber band into an 18–20 cm single length and

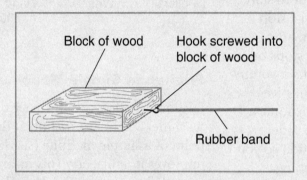

Block of wood

Hook screwed into block of wood

Rubber band

tie it to the hook in the block of wood as shown in the diagram.

Place the block of wood on one of the surfaces. Stretch the rubber band slowly in an effort to pull the block of wood. Use the ruler to measure how much the rubber band stretches before the block of wood moves forward. Repeat the procedure several times on the same surface and average the measurements. Record your results. Place a small book on top of the block and repeat the procedure. Record your results. Repeat both procedures for each of the five surfaces and record the results.

Stretch rubber band until block moves

Place various surfaces under block

Ruler

Questions

1. Design a chart and enter all the data.
2. Why did the length of the stretched rubber band differ among the surfaces?
3. Which surface had the greatest friction?
4. Which surface had the least friction?
5. What is the relationship between the weight of an object and the amount of frictional force opposing its motion?

Activities

1. At the Internet site, Fear of Physics, *http://www. fearofphysics.com/Friction/friction.html*, there is an interactive friction activity. The activity allows you to select a vehicle, control the speed of the vehicle, set road conditions, and determine when the brakes are applied. The animated results illustrate how successful you are in stopping the vehicle behind other cars.

   The chart below shows the results of six test runs. You may also want to try some other combinations of data and record the results to help you answer the questions.

Vehicle	Speed	Conditions	Brakes Applied	Results
SUV	10 m/s	Dry	5 meters	Crash
SUV	10 m/s	Dry	10 meters	Good Stop
SUV	20 m/s	Dry	10 meters	Crash
SUV	20 m/s	Dry	30 meters	Good Stop
SUV	20 m/s	Wet	30 meters	Crash
SUV	20 m/s	Wet	50 meters	Good Stop

a. How does the speed of a vehicle affect its ability to stop?

b. How do road conditions affect a vehicle's ability to stop?

c. How does the force of friction affect a vehicle's ability to stop?

2. Construct a model to demonstrate how friction can be reduced using ball bearings and oil. You need the following items to build the model: 1) a plastic jar lid, 2) a can from the pantry (the jar lid should be able to fit over the can), 3) 8–10 marbles, 4) a book, and 5) a few drops of oil.

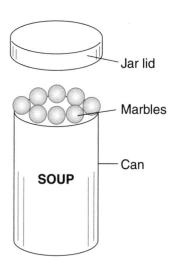

a. Place the jar lid over the top of the can and put the book on top of the jar lid. Spin the book on the jar lid. Describe and explain what happens.

b. Put some marbles under the lid and try spinning the book again. Describe and explain what happens.

c. Put a few drops of oil under the marbles and try spinning the book again. Describe and explain what happens.

Questions

1. When you push a penny and make it slide across a long table, it will eventually come to a stop. What force brings the penny to a stop?
   (1) gravity
   (2) friction
   (3) the air
   (4) upward force of the table

Base your answers to questions 2 and 3 on the following information and diagram.

A 50-newton block is being pulled along a metal surface by a string attached to a spring scale.

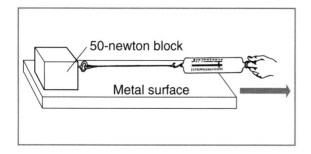

2. The force required to pull the block and make it slide along the surface is
   (1) greater than 50 newtons
   (2) less than 50 newtons
   (3) exactly 50 newtons
   (4) not possible to determine

3. How can the force of friction be decreased so that less effort is needed to move the block?
   (1) place sandpaper between the block and the metal surface
   (2) place oil between the block and the metal surface
   (3) change the metal surface to rubber
   (4) change the metal surface to wood

4. Friction acts on a bicycle to help make turns. Where does the friction occur that causes this?
   (1) in the seat
   (2) in the brakes
   (3) in the chain
   (4) between the tires and the road

5. What does friction do to the efficiency of a simple machine?
   (1) Friction does not affect the efficiency of a simple machine.
   (2) Friction increases the efficiency of a simple machine.
   (3) Friction decreases the efficiency of a simple machine.
   (4) Friction improves the efficiency of a simple machine to 100%.

Thinking and Analyzing

1. Friction is an important force in the operation of a car. Friction between the road and tires provides the force necessary to allow a car to start moving forward, to keep the tires from sliding on the road during turns, and to help stop the car. Explain what would happen if there were no friction between the road and the tires when
   (a) a car tried to start moving forward
   (b) a car tried to make a right-hand turn
   (c) a car tried to stop

2. Mary and her friends went sledding at the local park on a snowy winter day. During the first run down the hill, Mary realized that her sled went much too fast. How could Mary slow her sled when going down the hill?

2.7 What Is the Law of Conservation of Energy?

Objectives
Describe the Law of Conservation of Energy.

Describe different types of energy transformation.

Terms
Law of Conservation (kon-sur-VA-shan) **of Energy:** states that energy in the universe cannot be created nor destroyed; it can only be changed into another form

Law of Conservation of Matter: states that matter in the universe cannot be created nor destroyed

transformation (trans-fur-MA-shan) **of energy:** the changing of one form of energy into another form

Conservation of Energy

The **Law of Conservation of Energy** states that energy can be neither created nor destroyed. However, energy can be transformed from one form of energy into one or more other forms of energy. For example, when a match burns, you are changing chemical energy into light and heat energy.

This law is closely linked to a second law, the **Law of Conservation of Matter.** This law states that matter can be neither created nor destroyed.

In combination, these two laws state that neither energy nor matter can be created or destroyed, and energy and matter are "interchangeable." This means that energy can be changed into matter, and matter can be changed into energy. It also means that, from the beginning of time, the total amount of energy and matter in the universe has been constant and will remain constant in the future.

In our sun, matter is being changed into light and heat energy. To do this, a large amount of hydrogen is being changed into a lesser amount of helium. As the amount of matter in our sun decreases, the amount of energy released in the universe increases. It is much more difficult to give an example of energy being changed into matter. Under special laboratory conditions, scientists have been able to change energy into matter. This is not a common process under normal temperature and pressure conditions on Earth.

Energy Transformations

The **transformation of energy** occurs when one form of energy is changed into another

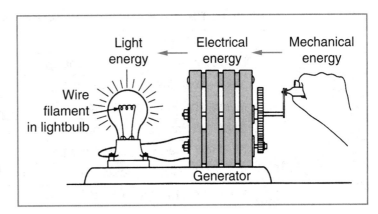

Figure 2.7-1. This hand-operated generator transforms mechanical energy into electrical energy, which is then transformed into light energy.

form of energy. Previously we discussed how different forms of energy could be changed into mechanical energy (see Lesson 2.5). But, energy transformation takes place in many other ways, too.

For instance, when you make toast in the morning, electrical energy is changed into heat energy that toasts the bread. At school, when the bell rings between classes, electrical energy is transformed into sound energy, a type of mechanical energy. And at night, when you turn on a reading light, electrical energy is changed into light energy. Figure 2.7-1 shows two common energy transformations. *Stop and Think:* What other examples of energy transformations can you name? *Answer:* There are many.

Turn on a television and electrical energy is turned into light and sound energy. Use a battery-operated toy and chemical energy is changed into mechanical energy.

During every energy transformation, the total amount of energy input is equal to the total amount of energy output. However, undesired forms of energy are also produced. For example, the purpose of a motor is to change electrical energy into mechanical energy. But some of the electrical energy is changed into undesired heat and sound energy. (See Figure 2.7-2.) Every energy transformation produces a small amount of heat energy that is released back into the universe.

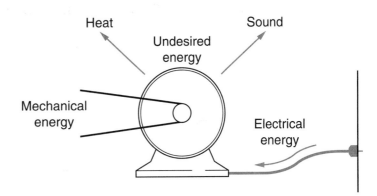

Figure 2.7-2. An electric motor transforms electrical energy into mechanical energy. Some undesired heat and sound energy is also produced.

Table 2.7-1   Energy Transformations on a School Bus

Ignited Gasoline	Chemical Energy → Heat Energy
Pistons	Heat Energy → Mechanical Energy
Rubbing Parts	Mechanical Energy → Heat Energy
Generator	Mechanical Energy → Electrical Energy
Battery	Chemical Energy → (Mechanical, Light, Sound) Energy
Headlights	Electrical Energy → Light Energy
Radio	Electrical Energy → Mechanical (Sound) Energy
Tire Friction	Mechanical Energy → Heat Energy

Flow of Energy

Our daily activities are filled with energy transformations. If we observe carefully, we can recognize the flow of energy from one form to another.

Almost all energy on Earth originates from the sun. Photosynthesis in plants changes light energy from the sun to chemical energy in fruits and vegetables. Our body changes the chemical energy in foods to chemical energy in our body. Muscles within our body change the chemical energy into mechanical energy in a swinging baseball bat. The mechanical energy in a swinging bat is changed to mechanical energy in a moving baseball. When the fielder's glove catches the ball, a small amount of heat energy is produced in the glove.

Energy transformations that take place on a school bus provide a more complex example of how energy flows within a system. Table 2.7-1 lists some of the many energy transformations that take place on a school bus.

Activities

1. Look around your house and find energy transformations. Reproduce the chart below and list the process or item where the energy transformation is taking place. For each process or item, state the energy transformation.

Process or Item	Energy Transformation

2. An incandescent lightbulb has an energy efficiency of 14 percent. That means that 14 percent of the electrical energy that goes into an incandescent lightbulb is used to produce light energy. The other 86 percent produces undesired energy. Place your hand near (**lightbulbs are hot—do not touch!**) an incandescent lightbulb that is turned on. What is the undesired energy that an incandescent lightbulb produces? Draw an illustration to show the energy transformation and efficiency of a lightbulb.

Interesting Facts About Energy Transformation on the Sun

The sun is a huge ball of glowing hydrogen gas. It has a diameter of about 1,400,000 km, or about 100 times the diameter of Earth. You could fit more than 1,000,000 Earths inside a hollow sun. The mass of the sun is greater than 330,000 times the mass of Earth. The sun produces energy from a process called fusion. Fusion is the process of combining 4 atoms of hydrogen to produce 1 atom of helium and large amounts of energy.

Every second, the sun converts 545,000 kg of hydrogen into 541,000 km of helium. The other 4000 kg is converted into energy. In other words, the sun loses 4000 kg of matter every second, because it is converted into light and heat energy. In one hour it loses 14,400,000 kg of matter, in one day it loses about 350,000,000 kg of matter, and in one year it loses about 128,000,000,000 kg of matter. The sun is about 5 billion years old, and despite the fact that it loses this much mass every year, it is expected to last another 5 billion years!

Questions

1. The diagram below shows a pinwheel rotating above a lit candle. The arrows indicate the direction of airflow.

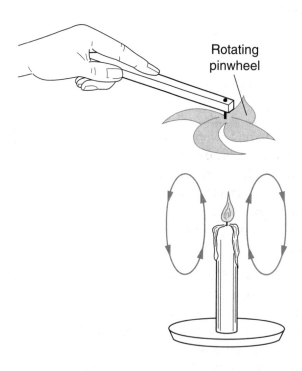

Which energy transformation is best shown in this diagram?
(1) heat to mechanical
(2) mechanical to light
(3) sound to heat
(4) heat to sound

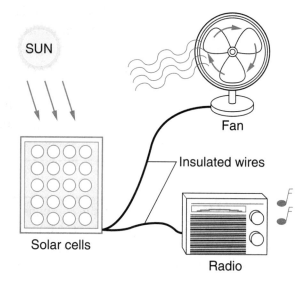

2. Which changes in energy are illustrated in the diagram above?
(1) electrical → sound → light and mechanical
(2) sound → mechanical → light and electrical
(3) mechanical → light → sound and electrical
(4) light → electrical → mechanical and sound

3. According to the Law of Conservation of Energy
(1) energy is being constantly created
(2) energy is being constantly destroyed
(3) energy is never created nor destroyed
(4) matter is being constantly created

4. A generator changes mechanical energy into electrical energy. This is an example of
(1) the creation of energy
(2) undesirable energy
(3) the destruction of energy
(4) an energy transformation

5. Which statement correctly describes what is happening in the sun?
   (1) The sun changes matter into energy.
   (2) The sun changes energy into matter.
   (3) The sun creates energy and creates matter.
   (4) The sun destroys energy and destroys matter.

Thinking and Analyzing

1. The diagram below shows the steps necessary to produce the energy needed to run a hair dryer.

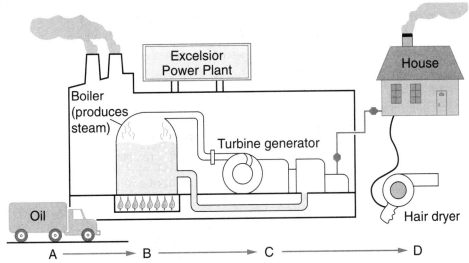

   As it moves from location A to location D in the diagram, the energy stored in the oil is being transformed.
   (a) What form of energy does oil contain?
   (b) What energy transformation is occurring in the boiler?
   (c) What energy transformation occurs in a turbine generator?
   (d) What energy transformation occurs in a hair dryer?

2. On a cold winter day, Mr. Clamis puts several logs into a fireplace and lights them with a match. After a few minutes, he has a fire heating his house. List the energy transformations necessary to get the energy from the sun to the energy in his house.

Review Questions

Term Identification

Each question below shows two terms from Chapter 2. One of the terms is defined.
(1) Choose the term that matches the definition.
(2) Describe how the two terms are different. Following each term is the section (in parenthesis) where the description or definition of that term is found.

1. *Potential Energy (2.1) — Kinetic Energy (2.1)*
   Stored energy of an object because of its position or condition.

2. *Energy (2.1) — Work (2.1)*
   What is done when a force moves an object a specific distance.

3. *Heat Energy (2.1) — Radiant Energy (2.1)*
   A form of energy that moves in waves and can travel through space.

4. *Resistance (2.2) — Effort (2.2)*
   The force a machine has to overcome.

5. *Simple Machine (2.2) — Complex Machine (2.3)*
   A device that uses only one motion to transfer mechanical energy.

6. *Joule (2.4) — Newton (2.4)*
   A unit of force in the metric system (equals 0.22 lb).

Multiple Choice (Part 1)

Choose the response that best completes the sentence or answers the question.

1. Energy is the ability to
   (1) apply a force
   (2) do work
   (3) apply an effort
   (4) apply a resistance

2. The state of energy that an object has because it is moving is called
   (1) heat energy
   (2) sound energy
   (3) potential energy
   (4) kinetic energy

3. One way that machines make work easier is by
   (1) increasing the effort force applied
   (2) decreasing the effort force applied
   (3) increasing the resistance force
   (4) decreasing the resistance force

4. What three simple machines are being used in the diagram to help move bricks from position A to position B?

(1) inclined plane, wheel and axle, and lever

(2) inclined plane, pulley, and lever

(3) inclined plane, wheel and axle, and wedge

(4) inclined plane, wedge, and lever

5. Work equals force times distance (W = F x d). How is the amount of work affected if the force increases, and the distance of motion remains the same?
(1) the amount of work remain the same
(2) the amount of work increases
(3) the amount of work decreases
(4) the amount of work increases then decreases

6. Under actual conditions what is the relationship between work input and work output?
(1) work input and work output are equal
(2) work input is always greater than work output
(3) work output is always greater than work input
(4) work output is sometimes greater than work input

7. Mechanical energy is
(1) the energy associated with waves from the sun
(2) produced by the flow of electrons through a substance
(3) the energy stored in the bonding of atoms and molecules
(4) the energy of a moving object

8. Which item changes electrical energy into mechanical energy?
(1) toaster
(2) lamp
(3) vacuum cleaner
(4) coffee pot

9. The chain of a bicycle is greased in order to
(1) increase weight
(2) reduce air drag
(3) reduce friction
(4) increase resistance

10. Friction decreases the efficiency of a simple machine. This means
(1) a greater effort is necessary to overcome resistance
(2) less effort is necessary to overcome resistance
(3) a greater resistance is produced
(4) less resistance is produced

11. According to the Law of Conservation of Energy and the Law of Conservation of Matter
(1) energy and matter are interchangeable
(2) energy can be changed into matter, but matter cannot be changed into energy
(3) matter can be changed into energy, but energy cannot be changed into matter
(4) energy and matter are being destroyed

12. A roller coaster is a complex machine because it
(1) contains one simple machine
(2) contains two or more simple machines
(3) has more than two moving parts
(4) has motion

Thinking and Analyzing (Part 2)

1. Melissa pulls a pendulum to position A in the diagram and releases it.

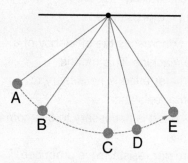

(a) At what position does the pendulum contain the maximum amount of potential energy?

(b) At what position does the pendulum contain the maximum amount of kinetic energy?

2. Base your answers on the diagram below and on your knowledge of science. The diagram shows two students ready to dive into a pool.

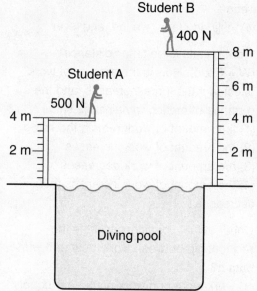

(a) Explain why student B has more potential energy than student A.

(b) Student A dives from the board into the water. Explain why student A's kinetic energy decreases as the student enters the water.

3. Mr. Jackson uses a pulley system to lift a 1000-newton box to a height of 4 meters.

(a) What is the work output?

    (1) 4000 N-m   (3) 250 N-m

    (2) 1000 N-m   (4) 500 N-m

(b) What is the work input?

    (1) 4000 N-m

    (2) greater than 4000 N-m

    (3) less than 4000 N-m

    (4) 0 N-m

(c) What is the efficiency of his pulley system?

    (1) 100 percent

    (2) greater than 100 percent

    (3) less than 100 percent

    (4) 0 percent

4. A simple machine is a device that uses a single motion to transfer mechanical energy. This means that mechanical energy goes into a simple machine and mechanical energy comes out of a simple machine. For each of the following items, identify the simple machine and describe the mechanical energy input and mechanical energy output.

(a) cutting bread with a knife

(b) pushing a wheelchair up a ramp

(c) turning a doorknob to open a door

(d) pulling a cord to raise window blinds

5. Almost all the energy on Earth comes from the sun. List the steps that trace the flow of energy from the sun to the mechanical energy you use when you walk.

6. Base your answers on the diagrams below and on your knowledge of science. The amount of force required to keep the same block of wood moving across a tabletop is shown in the diagrams.

(a) What change could be made in the setup in Diagram 1 that would require an increase in the amount of force necessary to move the block of wood?

(b) Explain why the round pencils in Diagram 2 decrease the amount of force necessary to move the block of wood.

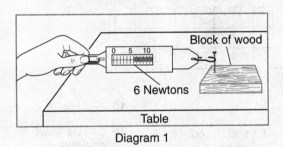

Diagram 1

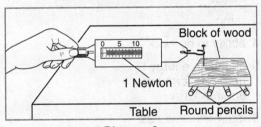

Diagram 2

7. Four machines that change electrical energy to mechanical energy were tested for fuel efficiency. The results of the test are shown in the chart below.

Test of Fuel Efficiencey		
Machine	Units of Electrical Energy Consumed	Equivalent Units of Mechanical Energy Produced
A	120	100
B	130	100
C	135	100
D	160	100

(a) Why is there a difference between the amount of energy consumed and the amount of energy produced?

(b) Which machine is most efficient? Explain why.

(c) Which machine is least efficient? Explain why.

Chapter Puzzle (*Hint:* The words in the puzzle are terms used in the chapter.)

Across

1 energy produced by flowing electrons

4 application of a force

7 energy of moving object

9 ratio of work output to work input

11 lever turning point

13 _____ of Conservation of Energy

15 energy of vibrating molecules

16 stored energy

17 done when a force moves an object

18 _____ machine contains 2 or more simple machines

Down

2 used to reduce friction

3 ramp

4 ability to do work

5 changing energy from one form into another form

6 force machine has to overcome

8 machine that uses 1 motion to transfer mechanical energy

10 force that resists motion

12 crowbar is example

14 bicycle chain is example

Unit 2
Weather

Part I—Essential Question

This unit focuses on the following essential question:

How do matter and energy interact to produce weather patterns?

The sun is Earth's primary source of energy. During the day, the sun warms the air and the ground. At night, the air and ground usually cool as Earth radiates heat back into space. Earth's land, water, and air constantly gain and lose heat, causing their temperatures to change. Temperature change is especially noticeable in the air, or atmosphere, around us.

When matter gains heat, it becomes less dense and expands. When matter loses heat, it becomes denser and contracts. In the atmosphere, warm, less dense air rises and denser air sinks. The rising and sinking of air on a small-scale produces local winds, land and sea breezes, and cumulus clouds. The rising and sinking of air on a large scale produces global wind belts, air masses, fronts, and storms.

Weather patterns are usually predictable if the weather elements are known. By gathering weather data such as temperature, air pressure, humidity, and winds, the meteorologist can predict future changes in the weather.

Part II—Chapter Overview

Understanding the interaction between the sun's energy and Earth's atmosphere provides a foundation for understanding weather.

Chapter 3 focuses on the properties of matter. There are three primary states of matter: solids, liquids, and gases. The addition of heat to matter and the removal of heat from matter causes the speed of the particles and the space between the particles of matter to change. Understanding how energy affects the density of matter is essential to understanding why air rises and sinks, and what causes weather changes.

Chapter 4 discusses heating and cooling events. It focuses on the three methods of heat transfer: radiation, convection, and conduction. All three methods transfer heat in the atmosphere. Heating and cooling events in the atmosphere are related to changes in air pressure, cloud formation, and winds.

Chapter 5 focuses on weather elements and events. All weather is caused by the unequal heating of Earth's surface by the sun's light and heat energy. Weather is usually described using four elements: temperature, air pressure, humidity, and winds. The water cycle, air masses, and storms are some weather patterns (events) produced by the exchange of energy in the atmosphere. The water cycle exchanges energy between air and water during the processes of evaporation and condensation. Clouds and precipitation are produced during the water cycle. Large air masses are produced when energy acts upon the atmosphere, hydrosphere, and lithosphere. Severe weather, such as stormy weather fronts, hurricanes, and tornadoes are produced during some of these energy exchanges in the atmosphere.

Chapter 3

Properties of Matter

Contents

The buoyancy of an object determines if it will float or sink in a fluid.

What Is This Chapter About?

Chemistry is the study of the properties, changes, and energy of matter.

In this chapter you will learn:

1. Matter is anything that has mass and takes up space.

2. The arrangement and motion of the particles in solids, liquids, and gases are different.

3. The density of an object determines whether it floats or sinks.

Career Planning: Chemist

Chemists work in different types of laboratories where products such as plastics, medicines, textiles, perfumes, foods, and fuels are made. They are involved in all aspects of the manufacturing process, including quality control. Many chemists are involved in testing the air, water, and soil in our environment. Other chemists produce chemicals used in agriculture to improve crops or eliminate insects. Some chemists, such as Josiah Willard Gibbs, pictured here on the postage stamp, are theoretical chemists. They

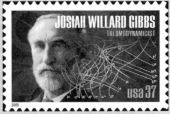

propose theories on the way nature works. His "phase rule" tells us that a solid, liquid, and gas can all exist at only one particular temperature and pressure. Gibbs's theories on energy changed the way we look at many chemical changes.

For more information on what chemists do, visit the following Web sites:

http://www.chem4kids.com/files/matter_intro.html We usually discuss three states of matter. At this Web site you can learn about two more.

http://www.dmturner.org/Teacher/Library/5thText/ChemPart1.htm Read about the scientific method, matter, mass, and volume.

http://www.angelfire.com/mo/matter/ A Web site that describes the general properties of matter: mass, volume, and weight.

3.1 What Is Matter?

Objectives

Define matter.

Distinguish between mass and volume.

Identify the units used to measure mass and volume.

Terms

matter: anything that has mass and takes up space

mass: a measure of the amount of matter in a sample

gram: a unit of mass

kilogram (KILL-oh-gram): a unit of mass equal to 1000 grams

triple beam balance: a tool used to measure mass

volume: the amount of space that a sample of matter takes up

liter (LEE-tuhr): a unit of volume

graduated cylinder (GRAJ-you-ay-ted SILL-in-duhr): a tool used to measure the volume of a liquid

Matter

Look around you. The objects you see, such as this book, your desk and chair, and the walls and ceiling, are all made of matter. The air you breathe and the water you drink are also made of matter. In fact, every solid, liquid, and gas is a form of matter.

Matter is defined as anything that has **mass** and takes up space. Mass is a measure of the total amount of matter in an object. We measure the mass of an object by comparing it to an object with a known mass. We can measure mass with an *equal-arm balance*, as shown in Figure 3.1-1. Figure 3.1-1 shows a container of water on

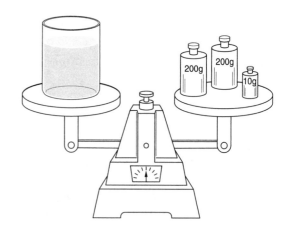

Figure 3.1-1. A balance is used to measure mass by comparing the object's mass to a known mass.

one side of an equal-arm balance. On the other side of the balance are three known masses labeled 200g, 200g, and 10g. Since the figure shows that the known masses and the container of water balance each other, we can conclude that the container of water has a mass of 410 grams.

In the laboratory, mass is often measured in units called **grams** (g). A penny has a mass of about 2.5 grams. The container of water in Figure 3.1-1 has a mass of 410 grams. When measuring the mass of larger objects, scientists use the **kilogram** (kg), which is equal to 1000 grams. The mass of an adult male weighing 187 pounds would be 85 kilograms. A mass of one kilogram has a weight of 2.2 pounds.

Measuring Mass

The equal-arm balance is just one of the tools that can be used to determine the mass of an object. Scientists also use a **triple beam balance** like the one shown in Figure 3.1-2. One advantage of a beam balance is that the standard masses are built into the balance. The mass of the object is found by adjusting the positions of the standard masses until the beam is perfectly balanced.

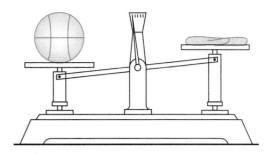

Figure 3.1-3. The inflated basketball is heavier and takes up more space than a deflated basketball because air is matter.

Air is matter even though we cannot see it. How could we prove that air has mass? Figure 3.1-3 compares the mass of a deflated basketball with the mass of a basketball filled with air. The difference in mass between the two is due to the air in the inflated basketball.

Also, notice that the air in the basketball takes up space and increases the size of the basketball. Air has mass and takes up space. The amount of space an object takes up is called its **volume**.

Measuring Volume

The volume of an object that has a regular shape can be found by using an

Figure 3.1-2. A triple beam balance is used to measure mass in units called grams.

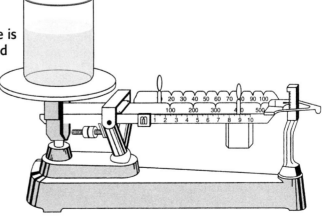

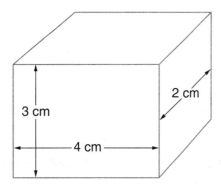

Figure 3.1-4. What is the volume of this box? To find the answer, multiply the measurements of length x width x height. The volume of this box is 24 cm³.

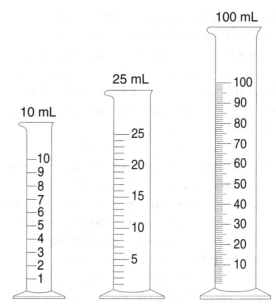

Figure 3.1-5. Graduated cylinders are used to measure the volume of a liquid. Measurements are in milliliters (mL).

equation or formula. For example, the volume of rectangular solids, such as the box shown in Figure 3.1-4, is calculated by multiplying length × width × height. The resulting unit of volume depends on the unit used to measure the box. In this example, the length, width, and height were measured with a centimeters (cm) ruler, so the volume is expressed in cubic centimeters (cm³ or cc).

A different unit of volume, the **liter** (L), is often used to express the volume of liquids and gases. A liter is just slightly larger than a quart. The basketball shown in Figure 3.1-3 would have a volume of about 7.5 liters when filled with air. The volume of a liquid can be measured using a **graduated cylinder** like those illustrated in Figure 3.1-5.

Energy

Is there anything that is **not** made of matter? Is there anything that has *no* mass and takes up *no* space? Figure 3.1-6

shows that shining a light on a balance has *no* effect on the measurement on the balance. This is because light is a form of energy. Energy is not matter, since it has no mass and no volume. Some other forms of energy are heat and sound.

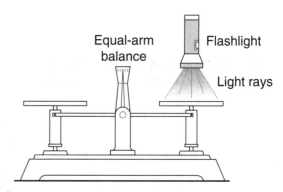

Figure 3.1-6. The balance is *not* affected by the light shining on it because light is *not* matter.

Activity

Stuff a paper towel in the bottom of a glass and invert it into a
pot of water. Why doesn't the towel get wet?

Questions

1. What does the diagram below show
about matter?

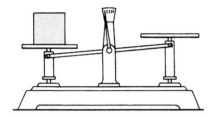

(1) Matter is made up of elements.
(2) Matter takes up space.
(3) Matter is a solid.
(4) Matter has mass.

2. Which is not an example of matter?
(1) water   (3) gold
(2) air   (4) sound

3. The amount of space an object occupies
is called its
(1) volume   (3) weight
(2) mass   (4) length

4. Which of the following is a unit of mass
in the metric system?
(1) liter   (3) kilogram
(2) pound   (4) cubic centimeter

5. The mass of a bowling ball is (see page
86)
(1) greatest at sea level
(2) lowest on the moon
(3) greatest on top of Mt. Everest
(4) the same at all of these locations

6. Which of the following correctly matches
a measuring device with the quantity it
measures?
(1) triple beam balance—length
(2) graduated cylinder—volume
(3) ruler—mass
(4) equal-arm balance—volume

Thinking and Analyzing

Describe what happens to the mass and
volume of an empty balloon as it is filled
with helium. Explain why this occurs.

Interesting Facts About Mass and Weight

People often confuse the terms *mass* and *weight*. Mass and weight actually measure different properties. You may have heard that astronauts experience weightlessness while in space. You have never heard that they experience "masslessness."

The mass of an object depends only on the amount of matter making up the object. The weight of an object is the force acting on the object due to gravity. In other words, weight measures how strongly gravity is pulling on the object. In space, astronauts can be so far from Earth that they no longer experience the pull of its gravity. Therefore, they are weightless.

Do you want to lose a lot of weight quickly? If so, move to the moon. On the moon, the force of gravity is one-sixth that of Earth, and therefore, objects weigh one-sixth as much. Unfortunately, your mass on the moon would be exactly the same as your mass on Earth (See Figures 3.1-7 and 3.1-8.) The true goal of a diet is not to lose weight, but to lose mass!

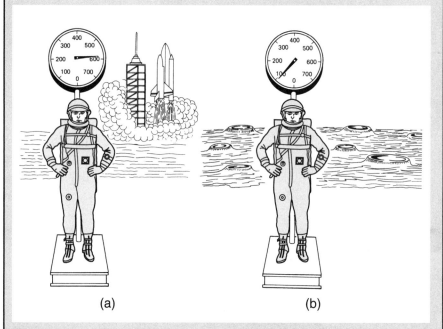

(a)   (b)

Figure 3.1-7. The weight of an astronaut is greater on Earth (*a*) than on the moon (*b*).

The unit of force, or weight, in the metric system is the newton. The unit of mass in the metric system is the kilogram. In the English system, pounds can be used for both mass and weight.

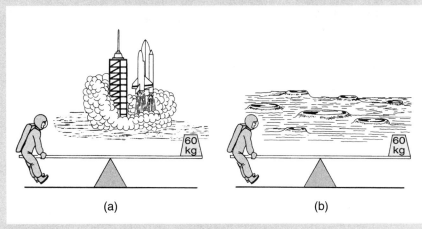

(a)   (b)

Figure 3.1-8. The mass of an astronaut is the same on Earth (*a*) and on the moon (*b*).

3.2

How Are Solids, Liquids, and Gases Different?

Objectives

Identify the differences among solids, liquids, and gases.

Draw diagrams illustrating some of these differences.

Terms

molecule (MAH- luh-kyool): the smallest particle of a substance that has the properties of that substance

substance: a pure form of matter that contains only one type of molecule

phase: the physical form of matter, such as gas, liquid, or solid; also called the **state**

solid: the phase of matter that has a definite shape and definite volume

liquid: the phase of matter that has no definite shape but has a definite volume

gas: the phase of matter that has no definite shape and no definite volume

fluid: a material that flows; a liquid or a gas

In the previous lesson, you learned that matter includes solids, liquids, and gases. In this lesson you will learn how these forms of matter are similar and how they are different. All forms of matter are made up of tiny particles called **molecules.** A molecule is defined as the smallest particle of a substance that is still that substance.

A **substance** is a form of matter that contains only one type of molecule. A drop of pure water is made up of 2 billion trillion (2,000,000,000,000,000,000,000) molecules of water—and nothing else. Because there is only one type of molecule, a substance is considered a *pure* form of matter. You probably know that the formula for the substance water is H_2O. A single H_2O unit is called a water molecule. Whether a

substance is solid, liquid, or gas depends on the arrangement of these molecules. The property that indicates whether a substance is solid, liquid, or gas is called its **phase** or its **state**.

Solids

Figure 3.2-1 is a model that compares the arrangements of the molecules in the three phases. Look at the arrangement of the molecules in the **solid**. The molecules are close together, and arranged in a regular pattern. They keep their positions, because they are strongly attracted to the molecules around them. Because the position of each molecule does not change, a solid keeps its shape. What would happen if the solid in

Figure 3.2-1. The three phases of matter—solid, liquid, and gas—differ in the arrangement of their molecules.

Solid

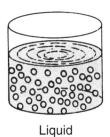
Liquid

Gas

the diagram were moved to a different container? It would look exactly the same. We say that a solid has a *definite shape*.

Liquids

In **liquids**, the molecules are usually farther apart than the molecules in solids. The molecules are *not* arranged in a regular pattern. They are able to move around within the liquid, and flow past each other. If the liquid shown in Figure 3.2-1 is poured into a tall narrow container, the liquid will change its shape. The liquid has no definite shape. It takes the shape of its container. The molecules in the liquid are attracted to each other, but not as strongly as the molecules in a solid.

Gases

In **gases**, the molecules are much farther apart than they are in liquids. Because the gas molecules have almost no attraction for each other, they can move anywhere within the container. Gases have no definite shape. Like liquids, gases are fluids, and take the shape of their container. A **fluid** is a material that can flow or move from place to place. Unlike liquids, however, gas molecules fill the container into which they are placed. Gases have no definite volume. If we put the

same amount of gas into a larger container, the gas moves everywhere inside that container and soon fills it up as well. The volume of a gas is defined as equaling the volume of the container.

Comparing the Three Phases

If you pour a quart of milk into a one-gallon jug, what is the volume of the milk? Of course it is still one quart. Liquids have a definite volume. The volume of a liquid depends on how much liquid there is, and *not* upon the size of the container. Solids have a definite volume as well. Only gases take the volumes of their containers.

In Figure 3.2-1, the molecules of the solid do not appear to be moving. However, molecules are always moving, no matter what phase they are in. Solids, liquids, and gases differ in the *way* the molecules move. In solids, strong attractions hold the molecules in position. The particles move back and forth, or vibrate, without changing their positions. In liquids, the molecules can move past each other *and change position*. That is why liquids have no definite shape. The molecules in a liquid still have a strong enough attraction to each other to keep them close together. That is why liquids have a definite volume. In gases, the molecules have total freedom of motion.

Table 3.2-1. A Summary of the Properties of Solids, Liquids, and Gases

	SOLIDS	LIQUIDS	GASES
Definite shape?	Yes	No	No
Definite volume?	Yes	Yes	No
Fluid?	No	Yes	Yes
Distance between molecules	Small	Greater	Greatest
Attraction between molecules	Greatest	Smaller	Smallest

They move freely through their container, bouncing off the container walls, and off other molecules as well.

When gas molecules strike the walls of the container, they exert a pressure on the container. If more gas is added to a metal container, the volume of the gas stays the same, but the pressure of the gas increases. More molecules in the same space produce a greater pressure. Table 3.2-1 summarizes the properties of solids, liquids, and gases.

Interesting Facts About Phases of Matter

Three phases of matter may not be enough. Scientists have identified several types of matter that do not fit the definitions of solids, liquids, and gases. Two more phases of matter are glass and plasma.

Glass has a definite shape, but the particles are not arranged in a regular pattern. Most solids, as discussed in the next lesson, will change from solid to liquid at one particular temperature, called the *melting point*. When glass is heated, it gradually gets softer and softer, until it begins to flow freely. Is glass a solid or a liquid? Many scientists say "Neither." They identify *glass* as another phase of matter.

The molecules in a gas are neutral particles. This means that they have no electrical charge. However, under certain conditions, such as extremely high temperatures, the particles become charged, and the substance's properties change. This new phase of matter is called **plasma**. All stars, including our sun, consist of plasma. Plasmas can also be found in fluorescent lights and television screens

Activity

For more information on the states of matter, visit *http://www.chem4kids.com/files/matter_intro.html* on the Internet. Read through the "Overview," and click on "Next Stop on Site Tour" until you finish reading "Looking For A Gas." Test your knowledge by taking the quizzes at the end of each page.

Questions

1. Which phase(s) of matter is considered fluid?
   (1) gases only
   (2) liquids only
   (3) liquids and gases
   (4) solids and gases

2. The circles in the closed jars shown below represent particles of matter. Which jar most likely contains a solid?

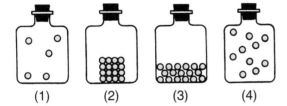

   (1)　　　(2)　　　(3)　　　(4)

3. Which phase of matter takes the volume of its container?
   (1) solid　　　　(3) gas
   (2) liquid　　　 (4) none of these

4. When the attraction between molecules are very strong, the substance is probably in the
   (1) solid phase　(3) gas phase
   (2) liquid phase　(4) none of these

5. Which of the following statements best describes a liquid?
   (1) A liquid has no definite shape and no definite volume.
   (2) A liquid has a definite shape and a definite volume.
   (3) A liquid has no definite shape and a definite volume.
   (4) A liquid has a definite shape and no definite volume.

Thinking and Analyzing

1. Describe how the molecules are arranged differently in solids, liquids, and gases.

2. Explain why molecules that have very little attraction fill their container.

3. Copy the flow chart at right into your notebook. Fill in the blanks with the correct phase of matter.

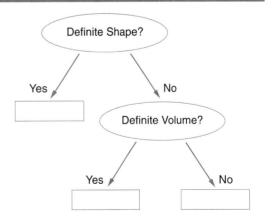

How Does Matter Change Phase?

Objectives

Identify the six types of phase change.

Describe the flow of energy during a phase change.

Define *melting point* and *boiling point*.

Give examples of phase changes.

Terms

boiling: a rapid phase change from liquid to gas that occurs at a particular temperature

melting: changing from solid to liquid

freezing: changing from liquid to solid

evaporating: a phase change from liquid to gas that occurs at the surface of a liquid at any temperature

condensing (kahn-DEN-sing): changing from gas to liquid

subliming (sub-LY-meeng): changing from solid to gas

depositing (dee-PAH-zeh-ting): changing from gas to solid

Boiling

Have you ever seen water boiling, as shown in Figure 3.3-1 on the next page? What happens if you continue to boil the water in a beaker for too long? The water disappears, and the heat may eventually crack the beaker. Where does the water go? Matter cannot be destroyed, but it can change from one form to another. When water **boils**, it turns into water vapor, an invisible gas. A change from liquid to gas is one example of a change in phase. In any phase change, the molecules remain the same, but the *arrangement* of the molecules changes. The formula for liquid water is the same as the formula for water vapor—H_2O.

Recall from the previous lesson that the distance between the molecules in a liquid is different from the distance between the molecules in a gas. The molecules of a gas are much farther apart than the molecules in a liquid. Heat is needed to separate the molecules. Because heat is required, boiling is an energy-absorbing process. All phase changes involve either the absorption or release of energy.

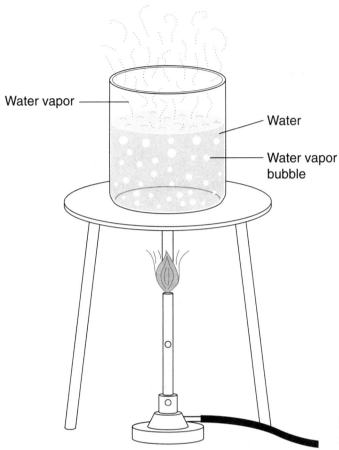

Water vapor

Water

Water vapor bubble

Figure 3.3-1. When water boils, it rapidly changes from a liquid to a gas.

How Liquids Boil

To understand the boiling process, we must look at the arrangement and motion of the molecules of the liquid. Molecules are always moving. An increase in temperature increases the speed of the molecules. Faster-moving molecules spread farther apart. If you heat the molecules of a liquid, they move faster and spread farther apart. Eventually, the molecules move so fast, and get so far apart, that they break away from each other and form a gas.

Changes of Phase

All phase changes can be explained in terms of molecular motion. The phase depends upon the speed of the molecules and the distance between them. At higher temperatures, molecules move faster, and spread out more. Higher temperatures tend to produce gases. Lower temperatures tend to produce solids. Therefore, a change in temperature can result in a change of phase.

Scientists have identified six kinds of phase changes. You are probably familiar with four of them: boiling, freezing,

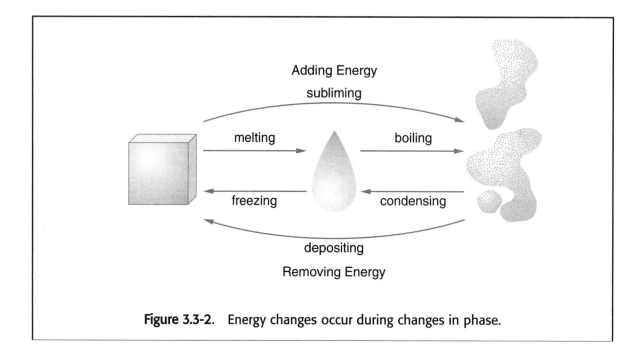

Figure 3.3-2. Energy changes occur during changes in phase.

melting, and condensing. Figure 3.3-2 identifies all six of these phase changes. Two of them, sublimation and deposition, are less frequently observed.

- **Melting** is the change from solid to liquid. To change into a liquid, the molecules in a solid must generally be moved farther apart, out of their fixed positions. Heat energy must be added to a substance to separate its molecules, so energy is *absorbed* during melting.
- **Freezing** is the *opposite* of melting. When a liquid freezes into a solid, the molecules move closer together and bond more tightly into fixed positions. The process of freezing *releases* energy.
- **Boiling** or **evaporation** is the change from liquid to gas. This change requires that the molecules of the liquid be separated even farther apart. Therefore, energy is *absorbed* when a liquid changes into a gas.

- **Condensation** is the change from gas to liquid. During condensation, the molecules of a gas move closer together to form a liquid, and energy is *released*.
- **Sublimation** is the change from solid directly to gas. During sublimation, the molecules of the solid move much farther apart to produce a gas. The *absorption* of energy is required to separate the molecules.
- **Deposition** is the opposite of sublimation. It is the change from gas directly to solid. The process of deposition *releases* energy.

Examples of Phase Changes

Phase change is one of the main events controlling our weather. Water vapor enters the atmosphere through evaporation. Water vapor in the air condenses to form dew, fog, and clouds. The role of these changes in

Activity

Take a glass and fill it halfway with ice. Leave it out for at least 30 minutes. Can you identify *three* phase changes that have occurred? Two are happening inside the glass, while the third is happening outside the glass. One of the phase changes inside the glass is easy to observe, while the other is not.

---

weather is discussed in Lesson 5.5. Have you ever seen dry ice? It is used in movies to create fog and scary drinks that appear to have white smoke coming out of them. Dry ice is also used to keep food cold during shipping. **Stop and Think:** Why is it called *"dry"* ice? **Answer:** Dry ice is actually solid carbon dioxide. Carbon dioxide changes directly from a solid to a gas without ever becoming a liquid. Carbon dioxide *sublimes*.

Table 3.3-1 lists some common phase changes involving water.

The Difference Between Boiling and Evaporation

Notice that the change from liquid to gas has two different names: boiling and evaporation. How are these phase changes different? Have you ever watched water boil? Not very exciting! But watching water evaporate, although difficult to see, would be even less exciting! When water boils, bubbles of water vapor form throughout the liquid. They rise to the

Table 3.3-1. Examples of Water Phase Changes

Phase Change	Name	Heat Flow	Examples
Liquid to gas	Boiling or evaporating	Water absorbs heat energy	Puddle evaporation; water boiling
Gas to liquid	Condensing	Heat energy is released	Cloud condensation; water condensing on cold window or mirror
Solid to liquid	Melting	Water absorbs heat energy	Snow melting on street; ice melting in soda
Liquid to solid	Freezing	Heat energy is released	Ice cubes forming in freezer; ice forming on lake surface
Solid to gas	Subliming	Ice absorbs heat energy	Snow disappearing even though the temperature remains below freezing
Gas to solid	Depositing	Heat energy is released	Frost forming in a freezer; frost forming on a car window

surface and escape into the air. When water evaporates, however, only the molecules at the surface are able to escape. Evaporation is a much slower process than boiling. Boiling occurs only at a particular temperature, the *boiling point*, but evaporation occurs at any temperature.

Interesting Facts About Phase Change

It's so cold that you can see your breath! Have you noticed that on a cold winter day you seem to be breathing out smoke? What you are actually breathing out is water! When you exhale, your breath contains water vapor, which is invisible. When the air is cold enough, the water vapor condenses and forms tiny droplets of liquid water, which are visible. You can observe the same phase change if you exhale on a mirror. In really cold weather, your breath's vapor may deposit as ice. Men with moustaches may notice small icicles forming on their moustaches.

Why does the bathroom mirror fog up when you take a hot shower? The hot water evaporates quickly, filling the bathroom air with water vapor. When the moist air hits the cold surface of the mirror, some of the vapor condenses back to liquid water, fogging the mirror. The amount of water vapor in the atmosphere is called the *humidity*. Humidity is discussed in Lesson 5.4.

Questions

1. Condensation refers to a phase change from
   (1) solid to liquid   (3) liquid to gas
   (2) liquid to solid   (4) gas to liquid

2. Which is an energy-*absorbing* process?
   (1) boiling   (3) condensation
   (2) freezing   (4) deposition

3. Sublimation is a change from
   (1) solid to liquid   (3) solid to gas
   (2) liquid to gas   (4) gas to liquid

4. A hair dryer works on the principle that an increase in temperature increases the rate of
   (1) condensation   (3) evaporation
   (2) melting   (4) freezing

Thinking and Analyzing

1. State the name of each of the phase changes indicated by the three arrows in the figure below.

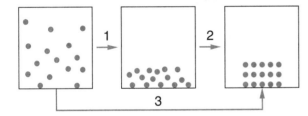

2. If you spill some water on the floor, and do not clean it up, it eventually disappears. What happens to the water?

3. After a snowstorm, even if the temperature remains too cold for the snow to melt, some of the snow gradually disappears. What is happening to the snow?

4. Isopropyl alcohol evaporates faster than water does at the same temperature. Which substance has a stronger attraction between its molecules? Explain.

3.4 How Do We Determine Melting and Boiling Points?

Objectives

Define melting point, boiling point, and freezing point.

Use a heating curve to determine the boiling point or melting point of a substance.

Terms

melting point: the temperature at which a solid turns to liquid

freezing point: the temperature at which a liquid turns to a solid

boiling point: the temperature at which a liquid bubbles and turns to a gas

heating curve: a graph used to determine the melting point or boiling point of a substance

Temperature and Phase Change

For every substance, the change in phase from solid to liquid occurs at a specific temperature called its **melting point**. The melting point of ice (the solid form of water) is 0°C. The temperature at which a liquid freezes into a solid is called its **freezing point**. The freezing point of water is also 0°C. *The freezing point and melting point of a substance are always the same temperature.* A substance will be a solid whenever its temperature is below its freezing point.

The temperature at which a liquid boils and changes rapidly to a gas is called its **boiling point**. The boiling point of water is 100°C. Water vapor, the gas form of water, also begins to condense into liquid water at 100°C.

While a substance is changing phase (see Lesson 3.3), its temperature remains

constant. For example, while you are boiling water, the temperature of the water remains at 100°C even though you are constantly supplying heat. The heat is used to cause the change in phase rather than to increase the temperature of the boiling water.

The Heating Curve

Both melting and boiling occur at a constant temperature. During these phase changes, heat is used to separate the molecules, and not to increase the temperature. One method of determining melting and boiling points is to measure the temperature of a substance while you heat it. If the temperature is increasing, the phase cannot be changing. When you reach a phase change, the temperature remains constant. A graph of temperature

Figure 3.4-1. A heating curve for water starts with ice at a temperature below its melting point.

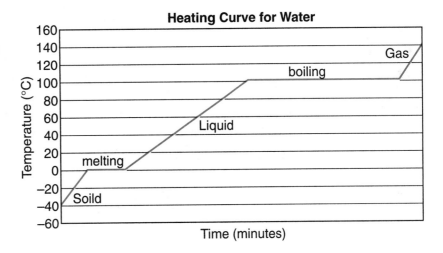

versus time, called a **heating curve**, is shown in Figure 3.4-1. As you can see, the graph appears flat during the phase changes. The melting and boiling points are the temperatures at which the graph flattens out. In the heating curve for water, the temperature stops increasing at 0°C, when melting begins. The temperature stays the same until all of the ice is melted. Once all of the ice is gone, the temperature increases until it reaches the boiling point.

A heating curve for a different substance has the same general shape, but different melting and boiling points. Figure 3.4-2 shows the heating curve of another substance.

Stop and Think: Can you find the melting point of this substance?

Answer: The temperature remains constant both at 80°C and at 220°C. The melting point must occur at a lower temperature than the boiling point. The melting point of this substance is 80°C.

The heating curve can be used to find the phase of a substance at any temperature.

Stop and Think: What is the phase of the substance in Figure 3.4-2 at a temperature of 160°C?

At any temperature below its freezing point, a substance is a solid. At a temperature above its boiling point, a substance is a gas. At a temperature between its boiling point and its freezing point, a substance is a liquid. *Answer:* At 160°C, this substance is a liquid.

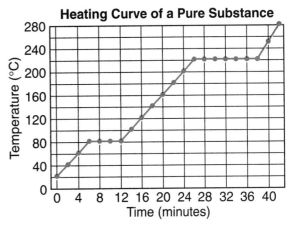

Figure 3.4-2. A heating curve for a pure unknown substance.

Interesting Facts About Boiling Points

A mixture of two liquids can be separated if the liquids have different boiling points. You heat the mixture until it just begins to boil. The liquid with the lower boiling point turns into a vapor and separates from the mixture first. You then cool the vapor in a different container and it condenses back to liquid. This process is called *distillation*. In a laboratory, distillation is often used to separate the parts of a mixture. The figure shows the equipment used to perform a distillation.

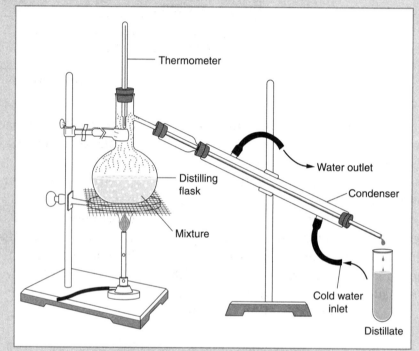

The equipment used in a laboratory distillation.

On a larger scale, this process is used to separate crude oil (also known as petroleum) into different products, such as gasoline and diesel fuel. This separation process occurs in an oil refinery. In this process, the temperatures at which the gases condense are used to separate them. The illustration shows some of the different substances distilled from petroleum, and the temperatures at which they condense.

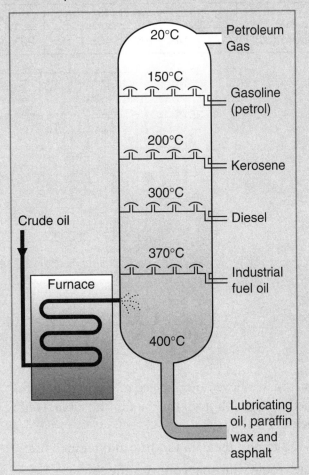

The components in crude oil are separated by distillation.

Skill Activity

Heat is applied at a constant rate to a solid substance under controlled conditions. The temperature of the substance was recorded every 3 minutes. The data are recorded in the table below:

Time (min)	0	3	6	9	12	15	18	21	24	27
Temperature (°C)	12°	14°	16°	16°	16°	20°	24°	28°	32°	36°

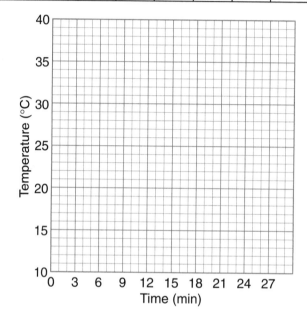

1. Copy the figure above onto a piece of graph paper and construct a line graph from the data in the table. Follow the steps below.

   A. Use Xs to plot the data for time and temperature.

   B. Draw a solid line that connects the Xs.

2. Provide an appropriate title for the graph.

3. According to your graph, what would the temperature of the substance be at 23 minutes?

4. If heat was added at a constant rate to the solid substance, why did the temperature remain at 16° for approximately 6 minutes?

5. Identify the independent and dependent variables in this experiment.

Questions

1. The temperature at which a substance melts is the same as the temperature at which it
   (1) boils
   (2) freezes
   (3) condenses
   (4) evaporates

2. The temperature at which a substance changes from liquid to gas is called its
   (1) boiling point
   (2) freezing point
   (3) melting point
   (4) sublimation point

3. A certain substance has a melting point of 52°C and a boiling point of 170°C. This substance is a liquid at a temperature of
   (1) 40°C
   (2) 50°C
   (3) 80°C
   (4) 180°C

4. A heating curve can be used to determine
   (1) only the melting point of a substance
   (2) only the freezing point of a substance
   (3) only the boiling point of a substance
   (4) the melting point, boiling point, and freezing point of a substance

Thinking and Analyzing

Base your answers to questions 1–3 on Figure 3.4-2 on page 99.

1. What is the boiling point of this substance?

2. What is the freezing point of this substance?

3. In what phase is the substance at 40°C?

4. Some solids like dry ice *sublime*. They turn directly from solid to gas. How would the heating curve of such a substance differ from the heating curves shown in this chapter?

3.5

How Is Density Calculated?

Objectives

Define density.

Calculate the mass, volume, or density using the formula: Density = mass divided by volume, or (D = m/v).

Term

density (DEHN-sih-tee): a quantity that compares the mass of an object to its volume

Defining Density

Why are airplanes made of aluminum, and fishing sinkers made of lead? You might answer that aluminum is a light metal, while lead is a heavy metal. Yet an aluminum airplane has a much larger mass than a lead fishing sinker. (When we say that lead is heavier than aluminum, we really mean that when we compare *equal volumes* of lead and aluminum, the lead piece is always heavier.) The quantity that compares the mass of an object to its size, or more specifically, to its volume, is called **density**.

Density is determined by dividing the mass of an object by its volume. (Density = mass/volume, or D = m/v.) While the mass and volume of a piece of metal depend on the size of the piece, the density is a property that depends on only the nature of the metal and its

temperature. Let's compare the densities of lead and aluminum.

Units of Density

At room temperature, the density of aluminum is 2.7 grams per cubic centimeter, which we abbreviate as 2.7 g/cm^3. Since density is mass divided by volume, the unit of density contains a mass unit, grams, divided by a volume unit, cubic centimeters. (Another unit of volume, the milliliter, abbreviated mL, is often used when measuring the volume of a liquid.) When we say that aluminum has a density of 2.7 g/cm^3, this means that a piece of aluminum with a volume of 1 cubic centimeter has a mass of 2.7 grams. The density of lead at room temperature is 11.3 g/cm^3. Lead is about four times as dense as aluminum.

Calculating Density

We can calculate the density of any object if we know both its mass and its volume. We use the equation D = m/v. For example, to determine the density of a sample of pure gold having a mass of 193 g and a volume of 10.0 cm³, we divide the mass by the volume (193 g/10.0 cm³). The density of the sample is 19.3 g/cm³.

Which is heavier, 15 grams of lead or 15 grams of aluminum? Of course, this is a trick question. Since both are 15 grams, they are equally heavy. We have already observed that lead has a greater density than aluminum. How is it possible for lead and aluminum to have different densities but have the same mass?

A 15-gram piece of aluminum is much larger in volume than a 15-gram piece of lead. In comparing objects of equal mass, the denser object is the one with the *smaller* volume. Figure 3.5-1 compares pieces of lead and aluminum of the same mass.

Stop and Think: The density of 10 cm³ of lead is 11.3 g/cm³. What is the density of 20 cm³ of lead? *Answer:* The answer is still 11.3 g/cm³! Table 3.5-1 gives the mass, volume, and density of several quantities of lead. As you can see, the density remains the

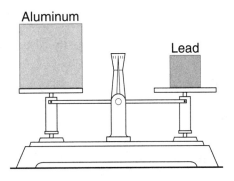

Figure 3.5-1. These metal cubes are of equal mass. The cube with the smaller volume has the greater density.

same no matter what size the metal is. The density of a material does not depend on its size. This makes density a very useful property for identifying substances.

Table 3.5-1. The mass, volume, and densities of several quantities of lead

Mass	Volume	Density (Mass/Volume)
113 g	10 cm³	11.3 g/cm³
226 g	20 cm³	11.3 g/cm³
1130 g	100 cm³	11.3 g/cm³

Activity

Use the Internet to find the density of the following metals:

Copper
Iron
Mercury
Zinc
Silver

One site you may want to use is *http://en.wikipedia.com* and search for each of the metals.

Why is density a very useful property when trying to identify gold?

Questions

1. Density is **best** defined as
   (1) mass times volume
   (2) mass divided by volume
   (3) volume divided by mass
   (4) mass plus volume

2. Calculate the density of an object that has a mass of 20 grams and a volume of 4.0 cm^3.
   (1) 0.20 g/cm^3 (3) 5.0 g/cm^3
   (2) 4.0 g/cm^3 (4) 80 g/cm^3

3. When iron is heated its volume increases. We can conclude that when iron is heated
   (1) as the volume increases, the density increases
   (2) as the volume increases, the density decreases
   (3) as the volume increases, the mass increases
   (4) as the volume increases, the mass decreases

4. A container of mercury has a volume of 10.0 cm^3 and a mass of 136 grams. What is the mass of 20.0 cm^3 of mercury? (*Hint:* remember that the density of mercury remains the same.)
   (1) 68 grams   (3) 272 grams
   (2) 136 grams   (4) 2720 grams

Thinking and Analyzing

1. The density of a quarter is 8.8 grams per cubic centimeter. What is the density of ten quarters? Explain

2. Why is density useful in identifying unknown substances, while mass alone is not?

3. Tinfoil and aluminum foil look very much alike. Aluminum has a density of 2.7 g/cm^3, while tin has a density of 7.3 g/cm^3. A chemist measures the volume of a sample of metal foil and finds it to be 5.0 cm^3. He then weighs it and finds its mass to be 36.5 grams. Is the sample of metal foil aluminum or tin? Explain your answer.

3.6

How Is Density Measured?

Objectives

To find the density of a solid or a liquid.

To measure volumes of solids by water displacement.

Terms

meniscus (muh-NIS-kus): the curve in the surface when a liquid, such as water, is placed in a glass container.

water displacement (dis-PLACE-ment): the replacement of a volume of water by the same volume of another substance that is placed in it

---

To determine the density of an object, you need two pieces of information—the mass and the volume. Lesson 3.1 describes how to find the mass of a solid using a triple beam balance. You learned how to find the volume of a liquid using a graduated cylinder (see Figure 3.1-5 on page 84). You also learned how to find the volume of a rectangular solid, using the formula *Volume = length × width × height*. What other skills are needed when measuring density? See below.

Measuring the Mass of a Liquid

We cannot determine the mass of a liquid by simply pouring it on the balance! We first find the mass of a suitable container for the liquid, such as the graduated cylinder shown in Figure 3.6-1.

We then add the liquid and find the mass of the liquid and container.

Subtracting the mass of the empty cylinder from the total mass of the cylinder and the liquid gives us the mass of the liquid.

Measuring the Volume of a Liquid

The volume of the liquid can be read directly from the cylinder. Figure 3.6-2 shows the proper way to read a graduated cylinder.

When measuring a volume of water in a glass container, you will notice a curve in the surface of the liquid. This curve is called a **meniscus**. The liquid should be read at the bottom of the curve, with the cylinder at eye level.

Measuring the Volume of a Solid Using Water Displacement

A graduated cylinder can be used to determine the volume of small solid objects, as well as the volume of liquids.

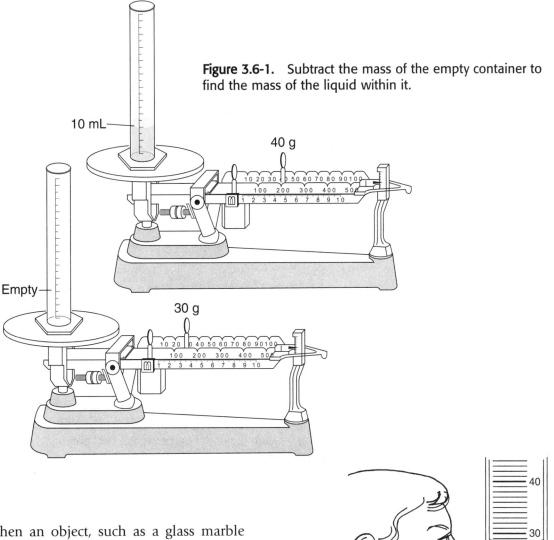

Figure 3.6-1. Subtract the mass of the empty container to find the mass of the liquid within it.

10 mL

40 g

Empty

30 g

When an object, such as a glass marble or small rock, is added to the water inside a cylinder, the water level rises. The change in the volume of the water is equal to the volume of the object placed in the water. An object that sinks in a liquid is said to have *displaced* an equal volume of water. A floating object must be pushed completely below the surface of the water so that its entire volume can be measured. This method of measuring the volume of a solid is called **water displacement**.

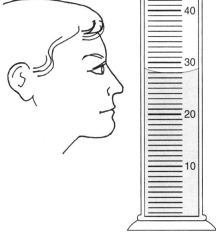

Figure 3.6-2. Read the volume of a liquid in a graduated cylinder at the bottom of the curve at eye level.

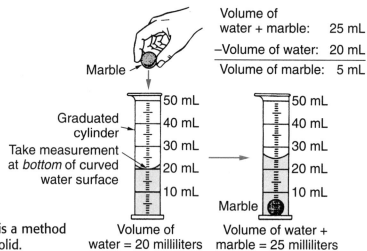

Volume of
water + marble: 25 mL
−Volume of water: 20 mL
Volume of marble: 5 mL

Figure 3.6-3. Water displacement is a method used to measure the volume of a solid.

Volume of water = 20 milliliters

Volume of water + marble = 25 milliliters

For example, in Figure 3.6-3, the graduated cylinder contains 20.0 milliliters (mL) of water. When the marble is placed in the cylinder, the water level rises to 25.0 mL. The change in the water's volume is 5.0 mL, so the volume of the marble is 5.0 mL (or 5.0 cm^3).

Interesting Facts About Water Displacement

The water displacement method for finding volume was discovered over 2200 years ago by Archimedes (ark-uh-MEE-deez), a Greek mathematician, astronomer, philosopher, physicist, and engineer. (We discuss his principle in the next chapter.) Archimedes was asked to determine if a crown made for the king contained pure gold. The king suspected that the goldsmith replaced some of the gold with silver and kept some gold for himself. Archimedes knew that gold was denser than silver. He could measure the mass of the crown with a balance, but how could he measure the volume? He found his answer while taking a bath! Archimedes noticed that as he entered the bath, the water overflowed. He jumped out of the bathtub and ran through the streets yelling "Eureka!" which means, "I've got it!" He realized that the amount of water an object displaces is equal to its volume. He used water displacement to measure the volume of the crown. Archimedes found that the volume of the crown was too large, and concluded that it was not made of pure gold. The lower density of the crown indicated that some of the gold had been replaced.

Questions

1. The diagram below indicates that the volume of the rock is

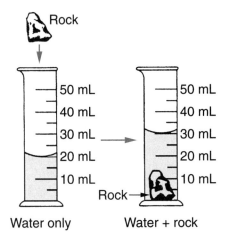

Water only　　Water + rock

(1) 10.0 mL　　(3) 30.0 mL
(2) 20.0 mL　　(4) 5.0 mL

2. A graduated cylinder contains 20.0 milliliters of water. If a stone with a volume of 12.0 milliliters and another stone with a volume of 5.0 milliliters are both placed into this cylinder, the water level will rise to show a total volume of
(1) 17.0 mL　　(3) 32.0 mL
(2) 25.0 mL　　(4) 37.0 mL

Questions 3 and 4 are based on the diagrams at the right.

3. Which tool will give a reading in milliliters?
(1) Tool A　　(3) Tool C
(2) Tool B　　(4) Tool D

4. Which tool can be used to determine the volume of a liquid?
(1) Tool A　　(3) Tool C
(2) Tool B　　(4) Tool D

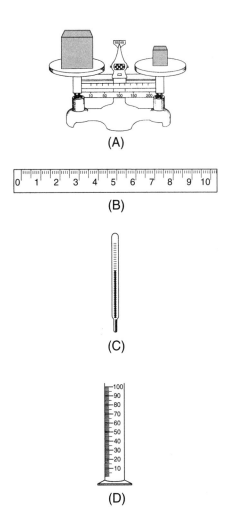

(A)

(B)

(C)

(D)

Thinking and Analyzing

For each of the units below, write the name of the property being measured and the tool used to measure it. The first one is done as an example.

Unit	Property	Tool
cm	Length	Ruler
cm³		
mL		
g		

3.7

Why Do Objects Float?

Objectives

To predict when objects float and when they sink.

To describe what happens to the weight of an object when placed in water.

Terms

buoyancy (BOY-uhn-see): the ability of an object to float in a fluid (liquid or gas)

Archimedes' (ark-uh-MEE-deez) **principle**: the weight lost by an object placed in a fluid is equal to the weight of the fluid that the object displaces

buoyant (BOY-uhnt) **force:** the upward force acting on an object placed in a fluid

---

Floating

A wooden log floats on water, while an iron nail sinks. Why? The explanation involves the densities of these materials. A material will float if it is less dense than the liquid in which it is placed. Iron is denser than water, so it sinks. Wood is less dense than water, so it floats. Water has a density of 1 g/cm³. Any object with a density greater than 1 g/cm³ will sink in water. The density of iron is 7.9 g/cm³. There are many different types of wood, but most have a density lower than that of water. Maple, for example, has a density of about 0.76 g/cm³.

When you picture an object floating in a liquid, you probably picture a *solid* object. Look at the liquids shown in Figure 3.7-1. The vegetable oil and vinegar in salad dressing form separate layers. What do you conclude about the density of vegetable oil compared with the density of vinegar? Since

the oil floats on top of the vinegar, the oil must be less dense. Oil is also less dense than water.

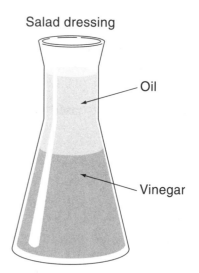

Figure 3.7-1. The less dense liquid (vegetable oil) floats on top of the more dense liquid (vinegar).

Crude oil, a dark, thick liquid, is used to make gasoline and other fuels. (See Lesson 3.4.) Sometimes, ships that are carrying crude oil are damaged, and the oil leaks out. The oil floats on the water, forming a thin sheet that can stretch hundreds of miles. The crude oil gets in the feathers of birds when they land on the water. Some oil spills have killed thousands of sea birds.

Floating in Air

Objects may float in a liquid, such as water. They may also float in a gas, such as air. Helium balloons float in air in much the same way that wood floats in water. The density of a helium balloon is much lower than the density of air. A balloon can move through the air, and a ship can move through the water. Because both gases and liquids are fluids, they permit objects to flow through them.

Some balloons, like the one in Figure 3.7-2, use hot air instead of helium. Since hot-air balloons float, hot air must be *less* dense than cold air. (See the Interesting Facts About Hot Air Balloons in Lesson 4.4.) When most materials are heated, they expand. This means that their volume increases while their mass stays the same. Since density is mass/volume, an increase in volume will cause a decrease in a material's density. In general, an increase in temperature causes a decrease in density.

Why Ice Floats

Which has a greater density, water or ice? You know that ice cubes float in water. Therefore, ice must be less dense than water. Water, unlike most substances, expands when it freezes. The larger volume of ice

Figure 3.7-2. A flame heats the air inside a hot-air balloon.

makes it less dense than water. Outdoor water pipes often burst in the winter when the water inside of them freezes and expands. A can of soda left in the freezer will eventually freeze and burst as well. The freezing liquid inside the can increases in volume until the can of soda breaks apart.

Buoyancy

Objects that float are said to be *buoyant*. The **buoyancy** of an object in a liquid depends on the density of the object in relation to the density of the fluid it is in. The less dense the object is, and the denser the fluid it is in, the greater the buoyancy of the object. Compare the way a cork floats on water with the way an ice cube floats. As shown in Figure 3.7-3, most of the cork remains out of the water, while most of the ice cube is below the surface. Cork is much less dense than ice, and so it is more buoyant.

Archimedes explained that when an object sinks in water, it displaces a volume

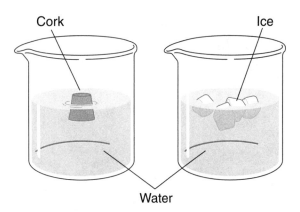

Cork   Ice

Water

Figure 3.7-3. A cork is more buoyant than an ice cube because it is less dense than the ice cube. The more buoyant object floats higher in the liquid.

of water equal to the volume of the object. (See the box feature in the previous lesson, "Interesting Facts About Water Displacement.") Archimedes also demonstrated that the object appears to lose weight when placed in water. You have probably noticed that it is easy to carry someone around in a swimming pool. In the water, people seem to weigh much less than their actual weight, even though their mass has not changed. How is this weight loss explained? Read below.

If you dip your hand into a fish tank, the water level in the tank rises. The rise in the water level is equal to the volume of water that your hand has displaced. In a sense, you have *lifted* that amount of water by pushing on it. You can feel the water pushing back. The push of the water is equal to the weight of the water that you lifted. The water is pushing your hand up, so your hand seems to be weightless.

Archimedes' Principle

When an object sinks in water, it displaces a volume of water equal to its own volume. The weight that the object appears to lose is equal to the weight of the water that it displaces. This is called **Archimedes' principle**. The upward force exerted by the water is called the **buoyant force**. The buoyant force is therefore equal to the weight of the water displaced by an object. Figure 3.7-4 shows how a rock loses weight when placed in water.

Some objects, when held under water, displace a weight of water greater than their own weight. When this happens, the buoyant force will be greater than the object's entire weight. The water is pushing

Figure 3.7-4. The weight lost by an object placed in water equals the weight of the water it displaces.

up harder than the object is pushing down. The object will float when it is released. Any object that is less dense than water will float in water, and appear to be weightless.

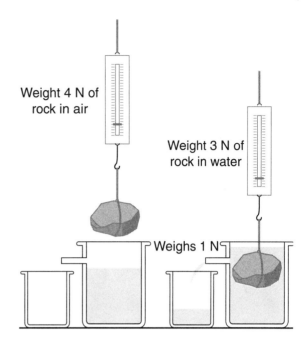

Weight 4 N of rock in air

Weight 3 N of rock in water

Weighs 1 N

Activity

The only difference between a can of ordinary cola and a can of diet cola of the same brand is the type of sweetener that is used. The cans have exactly the same volume. Submerge a can of cola and a can of diet cola in a sink filled with enough water to cover the cans. Let go of both cans, and observe any differences. Based on your observations, which can of soda contains a larger mass of sweetener? Justify your conclusion.

Interesting Facts About Buoyancy

Some objects float and some objects sink. A submarine, however, can do both. How is this possible? A submarine has special tanks called ballast tanks. These tanks can be filled with either air or water. As the ballast tanks are filled with water, the overall mass of the submarine increases. This increases the density of the submarine and it submerges or sinks. When the water is removed, and replaced with air, the mass decreases and the submarine floats. To stay at a particular level in the water without floating to the surface or sinking to the bottom, the density of the submarine must match the density of the water. This is called *neutral buoyancy*. For more information about submarines visit *http://www.yesmag.bc.ca/how_work/submarine.html.*

Questions

Questions 1 and 2 refer to the diagram below, which represents a beaker of ice water.

1. This diagram indicates that the ice is
   (1) less dense than water
   (2) more dense than water
   (3) colder than water
   (4) warmer than water

2. As water freezes to ice, its
   (1) volume decreases
   (2) mass decreases
   (3) volume increases
   (4) mass increases

Questions 3 and 4 refer to the diagram below, which shows the relative densities of some liquids and solids.

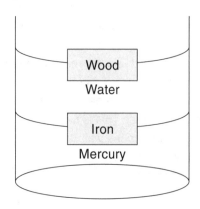

3. Which liquid in the diagram is least dense?
   (1) wood   (3) iron
   (2) water   (4) mercury

4. Which solid in the diagram is most dense?
   (1) wood   (3) iron
   (2) water   (4) mercury

5. The diagram below shows a tall container with four different liquids and their densities.

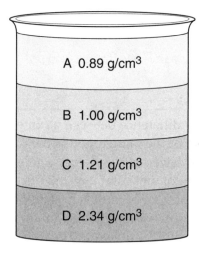

If a ball that has a density of 1.73 g/cm³ is placed in the container, where will the ball come to rest?
   (1) on top of liquid *A*
   (2) between liquids *B* and *C*
   (3) between liquids *C* and *D*
   (4) on the bottom of the container

Thinking and Analyzing

Base your answers to question 1 on the information and diagram below.

1. A student was given samples of four different liquids, *A, B, C,* and D. The student poured equal amounts of two different liquid samples into several test tubes. The results are shown in test tubes 1, 2, 3, and 4.

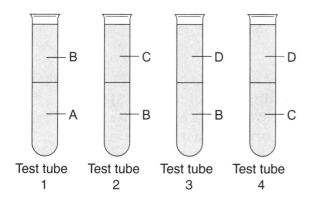

| Test tube 1 | Test tube 2 | Test tube 3 | Test tube 4 |

When equal amounts of liquids *A, B, C,* and *D* were placed into test tube 5, the liquids separated into four layers. A diagram of test tube 5 appears below. On the blank lines next to each layer, list the final order of the liquids, as they would appear in test tube 5.

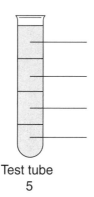

Test tube 5

2. Read the paragraph below and answer the question that follows it.

Iron has a density of 7.9 g/cm³, yet a ship made of iron can float in water. The ship is hollow, so its total volume is much larger than the volume of the iron alone. The volume of the ship is made up of the iron plus the air filling the space inside it. The mass of the air, however, is very small. The density of the ship, including the space inside it, is less than the density of water.

Why could a ship sink if it springs a leak?

Review Questions

Term Identification

Each question below shows two terms from Chapter 3. One of the terms is defined.
(1) Choose the term that matches the definition.
(2) Describe how the two terms are different. Following each term is the section (in parenthesis) where the description or definition of that term is found.

1. *Mass (3.1) — Volume (3.1)*
   The amount of space that a sample of matter takes up

2. *Kilogram (3.1) — Liter (3.1)*
   A unit of mass

3. *Graduated cylinder (3.1) — Triple beam balance (3.1)*
   A tool used to measure the volume of a liquid

4. *Molecule (3.2) — Phase (3.2)*
   The smallest particle of a substance that has the properties of that substance

5. *Solid (3.2) — Gas (3.2)*
   The phase of matter that has a definite shape and a definite volume

6. *Melting (3.3) — Freezing (3.3)*
   Changing from liquid to solid

7. *Boiling (3.3) — Condensing (3.3)*
   Changing from gas to liquid

8. *Meniscus (3.6) — Liquid (3.2)*
   The phase of matter that has no definite shape but has a definite volume

9. *Density (3.5) — Displacement (3.6)*
   The mass of an object divided by its volume

10. *Buoyancy (3.7) — Fluid (3.2)*
    A material that flows

Multiple Choice (Part I)

1. Which phase of matter has both a definite shape and a definite volume?
   (1) solid   (3) gas
   (2) liquid   (4) none of these

2. The diagram below shows milk being poured into a measuring cup.

   Which property of the milk can be directly measured using the cup?
   (1) mass   (3) solubility
   (2) density   (4) volume

3. Water vapor changes to liquid water during which process?
   (1) dissolving   (3) evaporation
   (2) melting   (4) condensation

4. A substance has a freezing point of –38°C and a boiling point of 356°C. At what temperature will this substance be in the liquid state?
   (1) –100°C   (3) 80°C
   (2) 400°C   (4) 375°C

5. The diagram below represents a glass container of water. A block has a density of 2.7 g/cm³. Which block in the diagram correctly shows what would happen when it is placed in the container of water?

   Container of Water

   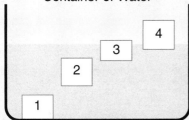

   Water Density = 1.0 g/cm³

   (1) Block 1   (3) Block 3
   (2) Block 2   (4) Block 4

6. A beaker of water is placed on a hot plate and heated until the water begins to boil. During this process water molecules
   (1) slow down and move closer together
   (2) speed up and move closer together
   (3) slow down and move farther apart
   (4) speed up and move farther apart

7. When water is heated from 25°C to 75°C it expands. As the water is heated its density
   (1) increases and its volume increases
   (2) increases and its volume decreases
   (3) decreases and its volume increases
   (4) decreases and its volume decreases

8. A block of wood has a mass of 9.0 g and a volume of 12.0 cm³. What is the density of the block of wood?
   (1) 1.3 g/cm³   (3) 3.0 g/cm³
   (2) 108 g/cm³   (4) 0.75 g/cm³

9. Which of the following is most likely to change when an astronaut visits an orbiting space station?
   (1) mass   (3) weight
   (2) volume   (4) height

Thinking and Analyzing (Part 2)

Base your answers to questions 1 and 2 on the diagram below and on your knowledge of science.

The diagram shows a phase change represented by letter A.

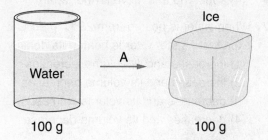

Water   100 g

A →

Ice

100 g

1. State the term for the phase change that occurs at A.

2. How does the motion of the molecules change during this process?

3. The drawings on the top row of the chart below represent water in its three phases (solid, liquid, and gas) in open containers. Complete this chart by filling in the answers that correspond to the drawing at the top of each column and the question in each row. Make sure you fill in an answer in every empty box.

	solid	liquid	gas
Does this phase of matter have a definite shape? Write Yes or No in each box.			
Does this phase of matter have a definite volume? Write Yes or No in each box.			
How do these phases rank in order or the relative speed of their particles? Rank them 1, 2, or 3 with 1 having the slowest particles and 3 having the fastest particles.			

4. Jorge wants to find the density of a glass marble. The tools used by Jorge to measure the glass marble are shown below. Use the information from the diagrams to calculate the density of the marble. Show your work.

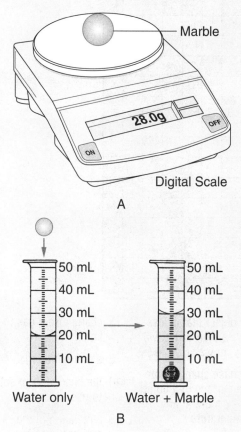

Marble

28.0g

OFF

ON

Digital Scale

A

50 mL
40 mL
30 mL
20 mL
10 mL

50 mL
40 mL
30 mL
20 mL
10 mL

Water only   Water + Marble

B

Base your answers to questions 5 through 8 on the information and chart below. Answer the question or complete the statement by filling in the blank. The chart below shows temperature readings recorded every minute while a substance was being heated at a constant rate. The material was a solid before heating and a hot liquid after 7 minutes of heating.

Time (min)	Temp (°C)
0	22
1	35
2	53
3	53
4	53
5	53
6	58
7	65

5. The temperature remained at 53°C for _____ minutes.

6. Was energy absorbed or released by the material during this time?

7. Identify the phase change that took place during this time.

8. Explain why the temperature did not change during this time.

Chapter Puzzle (*Hint:* The words in this puzzle are terms used in the chapter.)

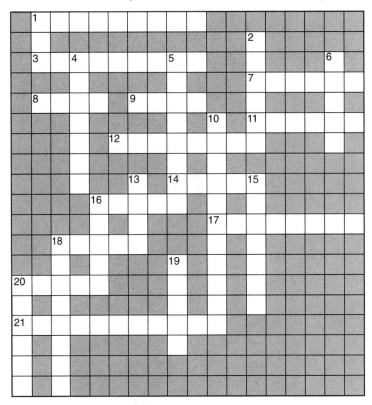

Across

1 a _____ cylinder is a tool used to measure the volume of a liquid

3 a pure form of matter

7 the phase of matter that has a definite volume but no definite shape

8 change from solid to liquid

9 an object that is denser than alcohol will _____ in alcohol

11 chemistry is the study of _____

12 the smallest part of a substance that has the properties of that substance

14 the phase of matter that has both definite shape and definite volume

16 the state of matter—solid, liquid or gas—is also called the _____

17 the curve in the surface when a liquid such as water is placed in a glass container

18 a liquid _____ when it changes rapidly from liquid to gas at a particular temperature

20 an object that is less dense than water will _____ in water

21 a phase change from the surface of a liquid that occurs at any temperature

Down

1 the phase of matter that has no definite shape and no definite volume

2 the amount of space a sample of matter occupies

4 a triple beam _____ is used to measure mass

5 change from gas to liquid

6 a unit of volume

10 the change from solid to gas

13 a measure of the amount of matter

15 the mass of an object divided by its volume

16 the temperature at which a substance melts is called its melting_____

18 the ability of an object to float in a fluid

19 a material that flows; a liquid or a gas

20 to change from liquid to solid

Chapter 4

Heating and Cooling Events

Contents

Strong convection currents produce violent storms
and transfer great amounts of heat in the atmosphere.

What Is This Chapter About?

Heating and cooling events constantly move
heat to and from the atmosphere. Water
plays a major role in this heat exchange.

In this chapter you will learn:

1. Heat is a form of energy produced by
   vibrating molecules in a substance.

2. Temperature is a measure of the average
   molecular motion (vibrating molecules)
   of a substance.

3. Radiation is the transfer of heat in the
   form of waves. Radiation travels from the
   sun to Earth.

4. Conduction is the transfer of heat from
   molecule to molecule.

5. Convection is the transfer of heat by the
   flowing action within a liquid or gas.

6. Heat is constantly added to and taken
   from the air by water phase changes.

7. Rising air expands because the
   surrounding air pressure decreases. This
   causes the rising air to become cooler.

Career Planning: Airplane Pilot

Have you ever thought of flying an airplane?
Pilots fly airplanes for either commercial or
private use. Commercial pilots transport
passengers and cargo. Other pilots test
aircraft and dust crops using chemicals. Pilots
are highly trained in their knowledge of
flying and their aircraft. They must also know
how to plan safe air routes and avoid severe
storms. In 1927, Charles Lindbergh became
famous for being the first pilot to fly solo
across the Atlantic Ocean. As a pioneer in
aviation, he provided an inspiration for
many pilots during the 1930s. Since then,
pilots have taken airplanes faster and farther
than ever imagined back in 1927.

Internet Sites:

*http://asd-www.larc.nasa.gov/SCOOL/cldchart
.html* Learn how to identify clouds from photos.
Links to cloud identification charts are also
provided.

http://www.sciencenetlinks.com/index.cfm
Explores high interest science topics with lessons
and interactive activities and contains links and
resources to other science sites.

4.1

What Is the Difference Between Heat and Temperature?

Objectives
Explain the difference between heat and temperature.

Describe heat gain and heat loss between two substances.

Terms
heat: a form of energy produced by vibrating molecules in a substance

kinetic (kih-NEHT-ihk) **molecular** (muh-LE-kyuh-ler) **theory:** a theory that describes matter as containing molecules, which are in constant motion (vibrating)

temperature: a measure of the average molecular motion of the molecules of a substance; the greater the average molecular motion, the greater the temperature

Kinetic Molecular Theory

Heat is a form of energy produced by vibrating molecules. The **kinetic molecular theory** of matter states that matter is composed of tiny particles called molecules, and that these molecules are constantly vibrating. Since vibrating molecules produce heat, then all matter must contain heat.

The kinetic molecular theory also explains that the distance between molecules is related to their attractive forces and the energy in the molecules. The attractive forces of molecules tend to hold molecules together. The molecules in solids are strongly attracted to the other molecules surrounding them in the solid. That is why solids hold their shape.

Stop and Think: How would you describe the molecular attraction of the molecules in liquids and the molecules in gases?

Answer: Liquids have less molecular attraction than solids. Gases have the least molecular attraction.

Heat (energy) affects the strength of the attractive forces between molecules. When heat is *added* to a substance, the attractive forces become *weaker*. The molecules move farther apart. Their average vibrating motion *increases*. When heat is *removed* from a substance, the attractive forces become *stronger*. The molecules move closer together. Their average vibrating motion *decreases*. (See Figure 4.1-1.)

Heat and Temperature

Heat and temperature are related, but have different meanings. As stated above, heat is a form of energy that refers to vibrating

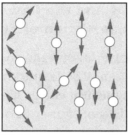 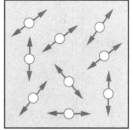 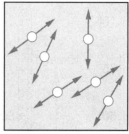

Figure 4.1-1.

(A) Molecules vibrating in a substance at 10°C

(B) Molecules vibrating in a substance at 30°C

(C) Molecules vibrating in a substance at 50°C

molecules. The amount of heat energy in a substance depends upon two things: (1) how fast the molecules vibrate, and (2) the amount, or mass, of a substance.

Temperature is a measure of the *average* molecular motion of a substance. By molecular motion, we are referring to vibrating molecules. However, temperature is *not* related to the mass of a substance.

The following examples should help distinguish between heat and temperature. If two beakers contain the same amount of water at different temperatures, the beaker with the higher temperature contains more heat. If two beakers contain different amounts of water at the same temperature,

the beaker containing the larger amount of water contains the greater amount of heat. (See Figure 4.1-2.)

Thermometer

A *thermometer* is an instrument that is used to measure temperature. Most substances expand when heated and contract when cooled. You can make a liquid thermometer by placing red-colored alcohol in a glass tube. If you add heat to the thermometer, the liquid expands and moves upward. Removing heat causes the liquid to contract and move downward. (See Figure 4.1-3 on the next page.) A coiled metal spring is

Figure 4.1-2. Beaker *B* contains more heat than beaker *A* even though they are both the same temperature, because there is more water in beaker *B*.

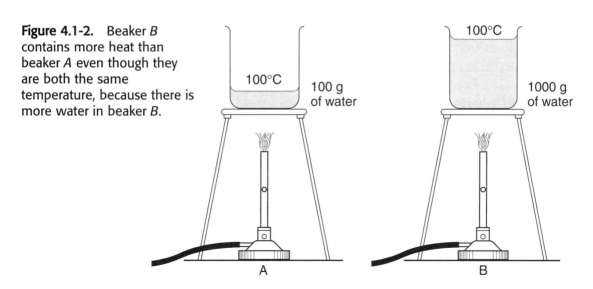

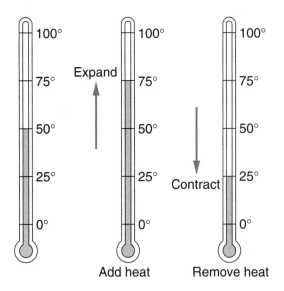

Figure 4.1-3. The liquid in a thermometer expands when heated and contracts when cooled.

sometimes used to make a thermometer. When heat is added the metal expands, and when heat is removed the metal contracts.

Two thermometer scales used to measure temperature are the Celsius (C) scale and the Fahrenheit (F) scale (See Figure 4.1-4). Each scale is divided into measurement units called degrees (°). The freezing and boiling points of water are known as fixed points because they are constant temperatures. Water freezes at 0°C, or 32°F, and it boils at 100°C, or 212°F. There are 100 degrees between the fixed points in the Celsius scale and 180 degrees between the fixed points in the Fahrenheit scale.

The Fahrenheit scale is often used for recording air temperature, but the Celsius scaled is preferred for scientific use.

Stop and Think: Which scale is usually used in weather forecasting?

Answer: The Fahrenheit scale is more commonly used in weather forecasting.

Heat and the Law of Conservation of Energy

Place a can of warm soda in the refrigerator. The can of soda quickly cools. It appears that the heat energy has been destroyed. But, the Law of Conservation of Energy (see Chapter 2) states that energy in the universe cannot be created or destroyed; it can only be changed into another form of energy. The heat in the can of soda was not destroyed nor was it changed into another form of energy. The heat moved from the can of soda to the air in the refrigerator, and eventually through the refrigerator system to the air in the room. Heat has the ability to move.

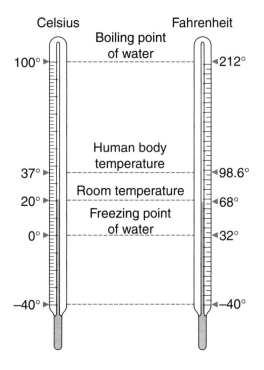

Figure 4.1-4. A comparison of the Celsius and Fahrenheit temperature scales showing some matching temperatures.

Transferring Heat

Heat moves from places or objects of higher temperature to places or objects of lower temperature when a difference in temperature exists. The tendency is for heat to move and equalize the temperatures. For example, place a cold glass of water on the table. Over a period of time, heat from the air will warm the water until the water and air are the same temperature.

Stop and Think: What happens to the heat in a cup of hot coffee that is left on the table?

Answer: The cup of hot coffee loses heat to the air, and eventually the coffee and air will be the same temperature.

The oceans and atmosphere transfer great amounts of heat. The movement of heat maintains a relatively consistent temperature range on Earth. (In Chapter 5 we will discuss how warm ocean currents carry heat from the equator toward the Poles.) In the Atlantic Ocean, the heat from the Gulf Stream affects the climate of the British Isles.

In the atmosphere, storms carry heat from warm regions to cooler regions. (In Chapter 5, we will also discuss how hurricanes carry heat from the tropics north to the United States or North Atlantic Ocean.) Winds also distribute heat in the lower atmosphere. Rising warm air carries heat from Earth's surface to higher elevations, and winds transfer heat across Earth's surface.

Heat is transferred by three different methods: conduction, convection, and radiation. In the next three lessons, you will learn about heat transfer in the atmosphere.

Activities

1. Heat moves from places or objects of higher temperature to places or objects of lower temperature. Try the two activities below and describe the results.

a) Hold a glass of ice water for about half a minute. Describe how your hand feels. Where did the heat in your hand go?

b) Explain why the ice in the glass did not melt.

c) Hold a cup of hot chocolate in your hand for about half a minute. Describe how your hand feels. Where did the heat come from?

2. Make a list of thermometers you have at home. Describe the use of each thermometer and tell what substance expands and contracts in each thermometer. Do you have any thermometers in which you cannot tell what expands and contracts?

SKILL EXERCISE—*Recording the Results of an Experiment*

One way of determining the relative amount of heat a liquid contains is to see how fast it melts ice. When you place an ice cube in warm water, heat moves from the water to the ice cube. The heat causes the ice cube to melt, or change into liquid water. The more heat in the water the faster the ice cube melts.

For this exercise you will need (a) three beakers (or cups), (b) a thermometer, (c) warm water (50–75°C), (d) three ice cubes about the same size, and (e) paper and pencil.

Put warm water in the three beakers. Completely fill beaker A, fill beaker B two-thirds full, and fill beaker C one-third full. Record the temperature of the water. Place an ice cube in each beaker of water. Observe the ice cubes and record the order in which they melt.

   a) Design a chart and place all your data in the chart.

   b) Compare the amount of heat in the three beakers at the beginning of the experiment. Explain why the amount of heat differed in the three beakers.

   c) In which beaker did the ice cube melt first? Explain why.

Interesting Facts About Temperature

Crickets are insects that make a chirping sound by rubbing their wings together. Only male crickets chirp. When it is warm they chirp more frequently and the number of chirps increases. When it is cool they chirp less frequently and the number of chirps decreases.

Did you know you can determine the temperature by the number of chirps a cricket makes? Count the number of chirps a cricket makes in 14 seconds Add 40 to this number to find the present temperature. If a cricket makes 32 chirps in 14 seconds, the temperature is (32 + 40) 72°F.

Questions

1. A form of energy produced by the vibrating motion of molecules is called
   (1) temperature   (3) heat
   (2) molecules   (4) sound

2. Two glasses of water are on a table. They both have the same temperature. Glass *A* contains 100 mL of water and glass *B* contains 200 mL of water. Which statement is true?

(1) Glass *A* has more heat than glass *B*.
(2) Glass *A* has less heat than glass *B*.
(3) Glass *A* has the same amount of heat as glass *B*.
(4) The amount of heat in both glasses constantly increases.

3. Removing heat from a substance causes the molecules in the substance to
   (1) stop moving
   (2) move faster
   (3) move slower
   (4) continue to move at the same rate

4. Heat is always transferred
   (1) from places and objects of higher temperature to places and objects of lower temperature
   (2) from places and objects of lower temperature to places and objects of higher temperature
   (3) from solids to liquids
   (4) from gases to solids

5. A weather thermometer measures the
   (1) radiation from the sun
   (2) reflected radiation
   (3) temperature of the air
   (4) temperature of the ground

6. When the temperature rises, the alcohol in a thermometer
   (1) contracts
   (2) expands
   (3) remains unchanged
   (4) changes from liquid to gas

Thinking and Analyzing

1. Which contains more heat and why?
   a) 500 mL of water in a teakettle at 100°C or 500 mL of water in a teakettle at 80°C
   b) 100 mL of soda in a glass at 10°C or 80 mL of soda in a glass at 10°C

2. When water *gains* heat, the area surrounding it *loses* heat. When water *loses* heat, the area surrounding it *gains* heat. For example, when you place hot coffee in a cool cup, the coffee loses heat and the cup gains heat. Explain in terms of heat gain and heat loss what happens when an ice cube melts in a dish.

3. A student made a thermometer using a flask, a 1-hole rubber stopper, some colored water, and a 30-cm glass tube.

When she wrapped her hands around the flask, the water rose up the tube. Why did the water rise up the tube?

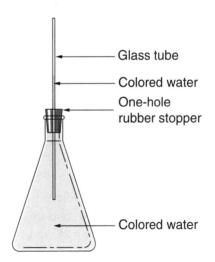

- Glass tube
- Colored water
- One-hole rubber stopper
- Colored water

How Is Heat Transferred by Radiation?

Objectives
Explain how heat is transferred by radiation.

Describe how radiation heats Earth.

Terms
radiation (ray-dee-AY-shuhn): the process of transferring energy in the form of waves

wavelength: the distance from the top of one wave to the top of the next wave

infrared (in-fruh-RED) **radiation:** a type of radiation that transfers heat through space

greenhouse effect: a process that allows short-wave radiation to pass through glass or the atmosphere, but prevents long-wave radiation from passing back out

---

Radiation

Radiation is the process of transferring energy in the form of waves. Heat waves and light waves are two forms of radiation we receive from the sun. Other examples of waves are radio waves, microwaves, and ultraviolet waves. Figure 4.2-1 shows how radiation travels in waves. One way in which waves differ is in **wavelength.** Wavelength is the distance from the top of one wave to the top of the next wave.

Earth receives different forms of radiation from the sun. The two most common forms of radiation we receive from the sun are *infrared radiation* and *light radiation.* Infrared radiation makes up 49 percent of all radiation from the sun, and light radiation makes up 43 percent. Other types of radiation make up 8 percent.

Infrared radiation is also called *heat radiation.* Light radiation, of course, makes things visible.

Two Forms of Radiation

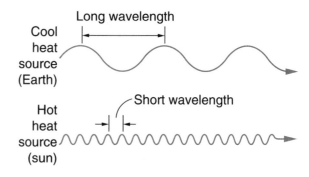

Figure 4.2-1. Radiation waves have a long wavelength from a cool heat source (Earth), and a short wavelength from a hot heat source (the sun).

All objects contain vibrating molecules. (See Lesson 4.1.) When molecules vibrate, they release heat radiation. In hot objects like the sun, the vibrating molecules produce short-wave heat radiation. In cooler objects like Earth's land and water, the vibrating molecules move slower and produce long-wave heat radiation. We can easily feel the short-wave heat radiation from the sun. We cannot easily feel the long-wave heat radiation from cooler objects. However, we can see long-wave heat radiation with special "night vision" glasses.

Heat radiation from the sun travels in a straight line outward at 300,000 km/sec. Heat radiation from the sun passes through space and our atmosphere to reach Earth. About 50 percent of the heat radiation that strikes Earth's surface is absorbed in the form of heat. Land is a good absorber of heat, water is a moderate absorber of heat, and the atmosphere is a poor absorber of heat.

Examples of Heat Radiation

Heat radiation and infrared radiation are the same thing. The sun is Earth's primary source of heat radiation. You can feel the sun's heat radiation on a sunny day.

There are other examples of heat radiation. Stand a few feet away from a campfire or the burning logs in a fireplace. The side of your body facing the fire gets warm, but your other side does not. The heat radiation from the fire travels in a straight line and is not able to heat the other side of your body. The heat from the fire reaches you by radiation.

An infrared lamp produces heat energy instead of light energy. The infrared energy from the bulb is used to treat some medical problems, such as sore muscles, arthritis,

and backache. The infrared waves travel through the skin and into the muscles. The muscles are warmed by radiation. An infrared lamp is also used to keep food hot in some restaurants and cafeterias.

The Greenhouse Effect

A car parked in the sun with all its windows closed gets extraordinarily hot. The hot sun emits short-wave radiation. The short-wave radiation easily passes through glass windows and heats the interior of a car. The dashboard and seats absorb the short-wave radiation. The objects in the car emit long-wave infrared radiation that cannot pass back out through the glass windows. The infrared radiation is trapped inside the car. (See Figure 4.2-2 on the next page.) If you have ever entered a closed car parked in the sun, you have experienced trapped heat radiation.

Trapped heat radiation is called the **greenhouse effect**. This process keeps greenhouses warm during cold weather. A greenhouse is built of glass and is used to grow plants. The glass allows the short-wave radiation to pass into the greenhouse, but does not allow the long-wave radiation to pass back out.

The greenhouse effect also keeps Earth warm. (See Figure 4.2-3 on the next page.) The atmosphere traps the heat energy much like the glass windows do in the car. The atmosphere allows the short-wave radiation from the sun to reach Earth, but allows only about 10 percent of the long-wave radiation to escape back into space. Water vapor and carbon dioxide in the atmosphere absorb most of the outgoing heat radiation. Without the greenhouse effect, all the heat radiation we receive from the sun in the daytime would be lost at night.

Figure 4.2-2. Car windows allow short-wave radiation from the sun to enter, but long-wave radiation cannot pass back out. The air in the car becomes very hot.

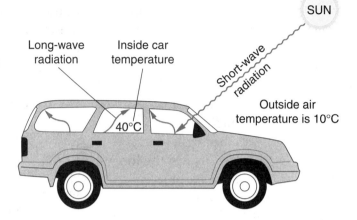

Figure 4.2-3. The greenhouse effect on Earth. Short-wave radiation reaches Earth, but the atmosphere traps most of the outgoing long-wave radiation.

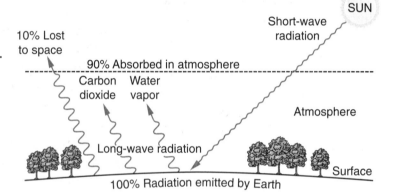

Activities

1. Try this activity on a cool sunny day.

a. Stand in the bright sun. How does your skin feel?

b. Stand in the shade of a large object. How does your skin feel? (Sometimes you need to move in and out of the sun to get the full effect of this activity.)

c. What caused the different feelings in your skin?

2. Try this activity in the early evening after a sunny, dry day. The sky should be clear. Extend your bare arm about 30 cm over a sidewalk or road surface. Turn your arm over several times so that your palm is up and then down. Hold your arm in place about 10 seconds each time you turn it over. You should feel one side of your arm getting warm.

a. Which side of your arm feels warm?

b. Why does your arm become warm?

Questions

1. Which process can transfer heat energy through space from the sun to Earth?
   (1) condensation   (3) sublimation
   (2) radiation   (4) evaporation

2. The two most common types of energy we receive from the sun are
   (1) Infrared and light energy
   (2) Infrared and sound energy
   (3) Infrared and microwave energy
   (4) light and sound energy

3. All objects emit infrared radiation. Which object emits the greatest amount of infrared radiation?
   (1) a copper rod at 10°C
   (2) a copper rod at 50°C
   (3) a copper rod at 100°C
   (4) a copper rod at 200°C

4. On a day when the outside temperature is 25°C, the temperature inside a closed parked car in the sun is 50°C. The cause of the higher temperature inside the car is
   (1) the greenhouse effect
   (2) light radiation
   (3) people sitting in the car
   (4) the car is not moving

5. The figure below shows how heat can be transferred. Each method is numbered. Based on the diagram, what method of heat transfer does number 1 indicate?
   (1) evaporation
   (2) deposition
   (3) radiation
   (4) condensation

6. The surface of Earth is heated by
   (1) heat radiation from the atmosphere
   (2) heat radiation from inside Earth
   (3) heat radiation from the sun
   (4) heat radiation from space

Thinking and Analyzing

1. Ms. Levy has three beakers of water on her desk. The first beaker contains 100 mL of ice. The second beaker contains 100 mL of water at room temperature. The third beaker contains 100 mL of boiling hot water. Compare the heat in the three beakers.
   a) Which beaker contains the most heat?
   b) Which beaker contains the least amount of heat?
   c) Explain how ice can contain heat.

2. Predict what would happen to the temperature on the night side of Earth if Earth's atmosphere did not contain water vapor and carbon dioxide.

SKILL EXERCISE—*Interpreting Data*

Dr. Ward did the following demonstration to show his students how land and water differ in absorbing and releasing heat radiation. He placed about 4 cm of water in an aluminum pan and 4 cm of black potting soil in another aluminum pan. He placed a thermometer to a depth of about 2 cm in both the water and potting soil. A lamp containing a 200-watt lightbulb was placed 30 centimeters above the pans, as shown in the figure below.

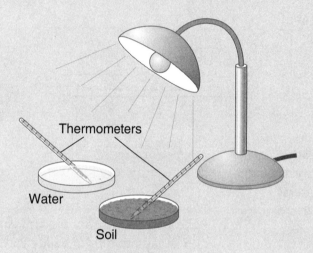

The starting temperature of each pan was recorded. When the lightbulb was turned on, students recorded the change in temperature every minute for ten minutes in each of the pans. When the light was turned off, the students continued recording the temperature change every minute for another ten minutes. The students recorded the data in charts as shown on the next page:

Light-On Chart—Heating Stage

Starting Temperature = 20°C

Time (minutes)	Water (°C)	Potting Soil (°C)
1	21	22
2	21	24
3	22	25
4	22	26
5	23	29
6	24	30
7	25	33
8	26	36
9	27	37
10	28	39

Light-Off Chart—Cooling Stage

Time (minutes)	Water (°C)	Potting Soil (°C)
1	28	38
2	28	37
3	28	36
4	27	35
5	27	34
6	26	32
7	26	29
8	25	27
9	25	24
10	24	22

Questions:

1. Which substance absorbed the most heat during the heating stage?

2. Which substance lost the most heat during the cooling stage?

3. How is this demonstration related to the land and water on Earth?

4. How does Earth get its heat in the daytime? What happens during the night to the heat radiation released from Earth?

4.3 How Is Heat Transferred by Conduction?

Objectives
Explain how heat is transferred by conduction.
Describe how conduction heats the atmosphere.

Terms
conduction (kuhn-DUHK-shuhn): the transfer of heat from one object to another object by direct contact

troposphere (TROH-puh-sfeer): lowest layer of the atmosphere where the temperature decreases with an increase in altitude

stratosphere (STRAT-uh-sfeer): second highest layer in the atmosphere where the temperature increases with an increase in altitude

mesosphere (MEHZ-uh-sfeer): third highest layer in the atmosphere where a decrease in temperature occurs with an increase in altitude

thermosphere (THUR-muh-sfeer): highest layer in the atmosphere

---

Conduction

Conduction is the transfer of heat by direct contact. Heat conduction can occur in solids, liquids, and gases. Solids are the best conductors of heat because the molecules in solids are closer together than the molecules in liquids or gases. Gases are the poorest conductors of heat because the molecules that make up gases are father apart than the molecules in solids or liquids.

Stop and Think: How do liquids compare as heat conductors to solids and gases?

Answer: Liquids are better heat conductors than gases, but not as good as solids.

Different substances conduct heat at different rates. Metal objects conduct heat well. (See Figure 4.3-1.) Pots and pans are commonly made of iron, copper, or aluminum. Table 4.3-1 lists the relative heat conductivity of some common substances. (Substances with higher numbers are better conductors than those with lower numbers.) When a metal pot is placed on a hot stove, heat is quickly distributed throughout the pot by the process of conduction. This allows the food in the pot to cook evenly. Some of the most advanced cooking utensils are made of ceramic materials. These solid materials withstand high temperatures and conduct heat rapidly.

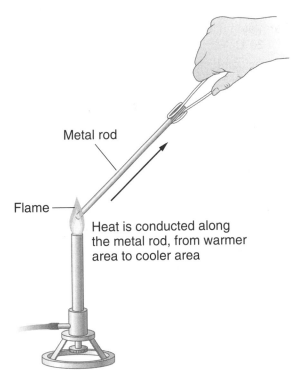

Figure 4. 3-1. Heating a metal rod is an example of heat transfer by conduction.

Table 4.3-1. Relative Conductivity of Some Common Substances

Substance	Measure of Conductivity	
Silver	420	High Conductivity
Copper	385	
Gold	320	
Aluminum	220	
Brass	100	
Steel	50	
Rock	1.7	
Glass	0.79	
Water	0.63	
Oak	0.147	
Sawdust	0.059	
Air	0.025	Low Conductivity
Vacuum	0	No Conductivity

Heating the Atmosphere by Conduction

The lowest layer of the atmosphere is called the **troposphere**. The troposphere extends from Earth's surface to an altitude of about 16 kilometers (10 miles). All weather takes place in the troposphere. The air in the troposphere that is in direct contact with Earth's surface is heated by conduction. The temperature of the air in the troposphere steadily decreases as the altitude increases.

In the previous lesson, you learned that the sun's radiation heats Earth's surface. The air in direct contact with Earth's surface gains or loses heat by conduction. When cool air moves over a warm land or water surface, heat is transferred into the air and

the air becomes warmer. If warm air moves over a cooler land or water surface, heat is transferred from the air to the surface and the air becomes cooler. Figure 4.3-2 on the next page shows how an air mass loses heat to the land by conduction.

Temperature Structure of the Atmosphere

The atmosphere consists of gases that are held to Earth's surface by gravity. About 97 percent of the gases are located in the lowest 30 km above the surface. Although there is no boundary to the upper limit of the atmosphere, the amount of gases thin out enough to suggest the top of the atmosphere to be about 150 km above the surface.

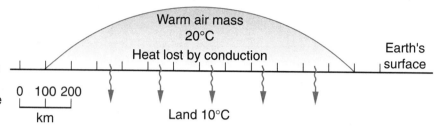

Figure 4.3-2. A warm air mass can lose heat energy to Earth's surface by conduction. The land becomes warmer and the air becomes cooler.

Scientists use temperature changes in the atmosphere as a way of identifying the four layers in the atmosphere. (See Figure 4.3-3.) We discussed the troposphere, the lowest layer, in the section above. The troposphere supports life with a sufficient amount of oxygen, water vapor, and carbon dioxide. As you rise in the troposphere, the temperature decreases. At the top of the troposphere,

there is a thin layer with a constant temperature of about –55°C called the *tropopause.*

The second layer, the **stratosphere**, extends to an altitude of 50 km and increases in temperature with an increase in altitude. The upper boundary of the stratosphere is the *stratopause*, which has a constant temperature of about 0°C.

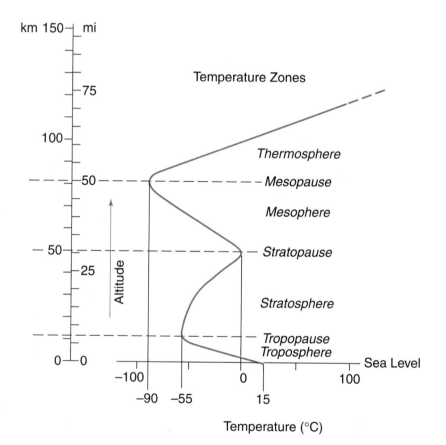

Figure 4.3-3. The structure of the atmosphere can be divided into layers based upon temperature zones.

The third layer is the **mesosphere**, which decreases in temperature with an increase in altitude. Above the mesosphere is the *mesopause,* a layer with a constant temperature of about −90°C.

The uppermost layer is the **thermosphere**. The thermosphere extends into space.

SKILL EXERCISE—*A Heat Experiment*

When a hot liquid is placed in a mug, the hot liquid rapidly loses heat. The heat is transferred from the liquid to the mug by conduction. You can place your hands around the mug and feel the heat in the mug. After the initial quick loss of heat from the liquid to the mug, the heat loss is slower and steadier. Eventually, the liquid and mug become the same temperature as the surrounding air. The following activity shows a method of collecting data to demonstrate how fast heat is transferred by conduction.

If you would like to duplicate this experiment you need 1) two coffee mugs, 2) enough hot water to fill the two mugs, 3) ice, 4) one digital thermometer to measure water temperature, and 5) a watch with a second hand.

a. Fill mug B with ice and water

b. Fill mug A with hot water. Immediately place the thermometer in the hot water and record the temperature. Continue recording the temperature every 30 seconds for 2 minutes.

c. Remove the ice water from mug B. Fill mug B with hot water. Place the thermometer in the water and record the temperature of the water. Continue recording the temperature every 30 seconds for 2 minutes.

The temperatures recorded for the above experiment were:

Mug A: starting temperature = 60°C, 30 seconds = 56.8°C, 60 seconds = 54.9°C, 90 seconds = 53.5°C, 120 seconds = 52.5°C

Mug B: starting temperature = 60°C, 30 seconds = 50.6°C, 60 seconds = 47.3°C, 90 seconds = 46.0°C, 120 seconds = 45.4°C.

1. Construct a chart using the data for mugs A and B.

2. What type of heat transfer does this activity demonstrate?

3. Which mug lost heat and which mug gained heat?

4. Why is the heat transfer greatest in the first 30 seconds?

5. Why was the heat transfer greater in mug B?

Questions

1. The transfer of heat by direct contact is called
   (1) condensation (3) evaporation
   (2) radiation   (4) conduction

2. Which of the following substances best conducts heat energy?
   (1) water   (3) copper
   (2) air   (4) wood

3. Heat conduction can occur in rocks, water, and air. Which statement correctly compares the ability of rocks, water, and air to conduct heat?
   (1) Air and rocks are better heat conductors than water.
   (2) Water and rocks are better heat conductors than air.
   (3) Air and water are better heat conductors than rocks.
   (4) Water is a better heat conductor than air and rocks.

4. The lower portion of the troposphere is heated by conduction. The conduction is between the bottom of the troposphere and the
   (1) sun
   (2) surface of Earth
   (3) interior of Earth
   (4) top of the atmosphere

5. Which statement correctly describes the troposphere?
   (1) as you go up in the troposphere the temperature increases
   (2) as you go up in the troposphere the temperature decreases
   (3) as you go up in the troposphere the temperature remains the same
   (4) as you go up in the troposphere you approach the region of conduction

6. Which action demonstrates heat transfer by conduction?
   (1) standing in sunlight
   (2) holding a cup of hot chocolate
   (3) standing in front of a fire
   (4) holding your hand over a candle

Thinking and Analyzing

1. As you rise in the troposphere the temperature decreases. The temperature decreases about 0.6°C for every 100 meters in altitude. Another way to state this is a decrease of 2.4°C for every 400 meters in altitude. The following lists the height of some well-known landmarks:

Mountain	Height
Mt. Marcy (New York State)	1600 m
Mt. Everest (Nepal-Tibet)	8710 m
Mt. McKinley (Alaska)	6010 m

Building	
Empire State Building (New York State)	375 m
Petronas Tower (Malaysia)	445 m

   a. Which mountain or building in the list has the coldest temperature at the top? Explain why.

   b. If the temperature at the base of the Empire State Building were 20°C, what would be the approximate temperature at the top of the building?

   c. If the temperature at the base of Mt. Marcy were 20° C, what would be the approximate temperature at the top of the mountain?

2. Use Figure 4.3-3 to answer the following questions:

   a. What property is used to define the layers of the atmosphere?

   b. In which atmospheric layer do you find clouds? Explain.

   c. Describe how the temperature changes as you increase your altitude in the stratosphere.

3. Mr. Johnson, the 6th grade science teacher, has three metal rods. Each rod is 30 cm long. The rods are made of steel, copper, and aluminum. Three students are asked to hold the rods at one end and place the other end in a Bunsen burner flame. Jose held the steel rod, Mike held the copper rod, and Alfie held the aluminum rod. Use Table 4.3-1 on page 137 to help answer the following questions:

   a. What method of heat transfer does this activity demonstrate?

   b. Predict which student felt the heat first, which student felt the heat second, and which student felt the heat last.

   c. Explain why the students felt the heat at different times

4.4

How Is Heat Transferred by Convection?

Objectives
Explain how heat is transferred by convection.
Describe an example of convection in the atmosphere.

Terms
convection (kuhn-VEHK-shuhn): process of heat transfer within a liquid or gas
convection current: movement of a liquid or gas caused by a change in temperature

---

Convection

Convection is the transfer of heat by a flowing action within a *fluid*. Both liquids and gases have the ability to flow, so they are called fluids. Adding heat to a fluid causes the molecules to move faster and farther apart, increasing its volume. This causes a decrease in density. In Chapter 3 you learned that a less dense substance in a fluid rises. The surrounding substance in the fluid is cooler and denser, so it sinks and takes the place of the rising fluid. The rising and sinking circulation within a fluid is called a **convection current**. Eventually, the heat generating the convection current is distributed equally throughout the fluid.

A simple demonstration shows how convection forms. Place a large beaker of water over a Bunsen burner. (See Figure 4.4-1.) The water directly above the flame becomes hot and begins to rise. The surrounding cooler water sinks to replace

the rising water. Place a few drops of food coloring in the water to observe the convection current.

Stop and Think: If the beaker is heated long enough, what do you think will

Water

Figure 4.4-1. A convection current can be produced in a beaker of water.

happen to the color of the water?

Answer: Eventually the convection current will mix the food coloring throughout the water.

Examples of Convection

Convection is used to distribute heat within a room. Typically, heat from a furnace enters a room along the base of a wall. As discussed in Lesson 4.3, the air near the heating unit gets warmer by conduction. The air rises above the heating unit. Cooler air from across the room flows to replace the rising air. A convection current is produced that evenly distributes the heat in the room. (See Figure 4.4-2.)

Hawks and eagles soar high in the sky looking for prey on the ground. When watching soaring birds, you soon learn that they seem to circle effortlessly without constantly flapping their wings. Soaring

birds actually stay within a convection current where warm air is rising. They can do this for hours.

Hot air balloons use the principle of convection to rise and fall. (See Interesting Facts About Hot Air Balloons at the end of this lesson.) Decreasing the density of the air in a balloon makes it rise. Increasing the density of the air in the balloon makes it fall.

Convection in the Atmosphere

The heating of Earth's surface depends to some extent on the nature of the surface. Some kinds of surfaces get hotter than others. For example, pavement and sand get much hotter than do grass and water. If you have ever walked barefoot on these surfaces on a warm sunny day you know this to be true. On a larger scale, the surfaces of oceans, forests, and deserts are

Figure 4.4-2. Home heating systems use convection currents to heat a room.

Baseboard heating

also affected differently by the sun. These surfaces, in turn, heat the air above them unequally, producing variations in temperature.

The convection of air produced by unequal heating of Earth's surface causes the formation of local surface winds. For example, Figure 4.4-3 shows the land heating faster than the water on a warm sunny day. The warmer air over the land rises, and the cooler air rushes in to take its place. The horizontal movement of air is called *wind*. Other types of winds will be discussed in Chapter 5.

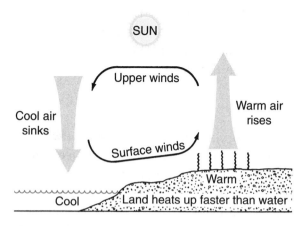

Figure 4.4-3. Unequal heating of Earth's surface causes convection and local winds to develop.

Activity

Place water in an ice cube tray. Place a few drops of food coloring in a couple of the ice compartments and place the tray in the freezer. You are making several colored ice cubes. Obtain a large glass beaker or a glass coffee pot and fill it with hot water (greater than 60°C). Place each colored ice cube in the water and record your observation.

SKILL EXERCISE—*Predicting Results*

Warm air is less dense than cold air. Therefore, warm air tends to rise and cold air tends to sink. To design an experiment to test this, you need cool and warm air, a way to control the flow of air, and several thermometers.

For instance, a small room with a window provides a means of controlling airflow. Three thermometers (*A*, *B*, and *C*) should be arranged on the wall as shown in the figure on the next page. On a day when the temperature outside is lower than the indoor temperature by at least 10°C, open the window to let in cold air and measure the temperature changes in the room with the thermometers.

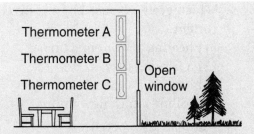

Answer the following questions about this experiment:

1. Which thermometer shows the first decrease in temperature? Why?

2. As the cold dense air enters the room, where does the warmer air in the room go? Explain.

3. What type of heat transfer does this experiment demonstrate?

Interesting Facts About Hot Air Balloons

A hot air balloon floating lazily across the sky captures your imagination. It leads to the question: How does a hot air balloon work and how is it controlled? The answer is based upon this scientific principle — hot air is less dense than cool air because it has less mass per unit of volume. Basically, a hot air balloon changes the density of the air in the balloon, causing the balloon to rise and fall.

A cubic foot of air usually weighs about 28 grams. Increase the temperature of the air 100°F and it loses 25 percent of its weight. The air now only weighs about 21 grams. It has a lifting power of about 7 grams per cubic foot. Although this is a small amount, large balloons contain many cubic feet of air and are capable of lifting hundreds of pounds.

A hot air balloon consists of three main parts: 1) a burner that heats the air in the balloon, 2) a basket that carries passengers, and 3) a balloon envelope that holds the warm air. When the burner is turned on, the air in the balloon is heated. The balloon will continue to rise as long as the air in the balloon is heated. To descend, warm air is released through an opening in the top of the balloon. This makes the air in the balloon cooler and denser, so the balloon descends. Direction is controlled by the surrounding wind. Wind direction differs at different altitudes. By rising or falling to the right altitude, the balloon can be placed in the wind that moves in the desired direction.

The highest flight in a hot air balloon was 34,466 meters (113,739 feet or about 21.5 miles) made in 1961. The first nonstop balloon flight around Earth (42,810 kilometers) was made in 1999.

Questions

1. Convection is a method of heat transfer. Convection takes place
   (1) only in solids   (3) only in gases
   (2) only in liquids (4) only in fluids

2. Local surface winds are caused by
   (1) unequal heating of Earth's surface
   (2) equal heating of Earth's surface
   (3) the revolution of Earth
   (4) the rotation of Earth

3. Which process is an example of convection?
   (1) Earth receiving energy from the sun
   (2) a metal rod being heated in a flame
   (3) warm air rising above a road surface
   (4) a frying pan being heated on a stove

4. Which event is caused by a convection current?
   (1) heating a room in a house
   (2) a flowing river
   (3) heating a spoon in hot water
   (4) rolling a rock downhill

5. Warm air rises because it is
   (1) less dense than cool air
   (2) more dense than cool air
   (3) the same density as cool air
   (4) heavier than cool air

Thinking and Analyzing

1. The figure below shows a chimney box built by a 6th grade science teacher. A burning candle was placed in the box as shown. The teacher then took a piece of thick rope and lit one end with a match. When the flame was extinguished, the rope produced smoke. The rope was held at positions A and B to show the flow of air in the box.

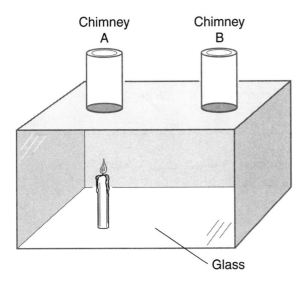

Chimney A   Chimney B

Glass

a. Describe the expected airflow at positions A and B.
b. What caused the airflow at position B?
c. What type of heat transfer does this demonstration show?

2. Every morning Mr. Sanchez, our principal, has a cup of coffee. He likes to add some cold milk to his coffee. Describe how Mr. Sanchez is producing a convection current when he puts cold milk in his cup of hot coffee.

4.5

How Does Water Exchange Heat in the Atmosphere?

Objectives

Describe the phase changes of water.

Describe the heat exchange during each phase change.

Explain how each phase change of water transfers heat in the atmosphere.

Terms

phase (fayz) **change:** changing from one phase of matter to another phase of matter; for example, in *melting*, a substance changes from a solid to a liquid

calorie (KA-luh-ree): the amount of heat energy needed to raise the temperature of 1 gram of water 1°C

Phase Changes of Water

In Chapter 2 you learned that there are three common phases of matter—solid, liquid, and gas—and each phase is associated with a specific molecular arrangement. You also learned there are six **phase changes**—melting, freezing, boiling (evaporation), condensation, sublimation, and deposition—and each phase change is caused by an exchange of heat energy.

Figure 4.5-1 on the next page shows the phase changes that occur in water as it gains and loses heat. Below its melting and freezing temperature (0°C), water is solid ice. Between 0° and 100°C, water is liquid. Above the boiling and condensation point temperature (100°C), water is a gas—water vapor.

The melting and freezing point of water is 0°C. The boiling and condensation point of water is 100°C. *Stop and Think:* What is the phase of water at 80°C? *Answer:* Water is in the liquid phase at 80°C.

Heat Exchange Between Water and Air

Heat energy is measured using a unit called a calorie. A **calorie** is the amount of heat energy needed to raise the temperature of 1 gram of water 1°C. Therefore, to change 5 grams of water from 40° to 50°C (a 10°C change), it takes 50 calories (5 × 10 = 50). *Stop and Think:* How much heat energy is removed from water when 10 grams of water decreases 2°C? *Answer:* When 10 grams of water decreases 2°C, the water loses 20 calories (10 × 2 = 20) of heat energy.

To weaken or strengthen molecular bonds during a phase change takes a much greater amount of heat. It takes 80 calories of heat to melt 1 gram of ice to water. By

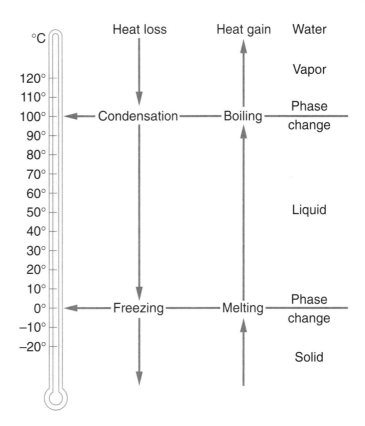

Figure 4.5-1. Phase changes that occur when water loses heat and when water gains heat.

contrast, it takes 540 calories to change 1 gram of liquid water to water vapor. Freezing removes 80 calories and condensation removes 540 calories from water. (See Figure 4.5-2 on the next page.)

When water gains heat something else must lose heat, and when water loses heat something else must gain heat. Commonly the heat exchange takes place between water and air. That is, when 1 gram of ice melts into liquid water, 80 calories of heat moves from the air to the ice in order to melt it.

Atmospheric Heat Exchange

Water phase changes in the lower atmosphere move large quantities of heat. The heat exchange occurs in the layer of air between the clouds and ground level. When water

gains heat, the surrounding air loses heat. When water loses heat, the surrounding air gains heat. Phase changes were introduced in Chapter 2. A review of the six phase changes, their heat exchange, and an example of where each occurs in the atmosphere is given below. Each phase change of water in the atmosphere produces a heat exchange.

Melting is the change from ice to liquid water. When snow and ice melt, the ice gains heat and the surrounding air loses heat. You commonly see this on a warm day after a snowstorm.

Freezing is the change from liquid water to ice. When a lake freezes, the water loses heat and the air gains heat.

Evaporation is the change of liquid water to water vapor. When a puddle of

Figure 4.5-2. Water gains calories when it changes from a solid to a liquid to a gas. Water loses calories when it changes from a gas to a liquid to a solid.

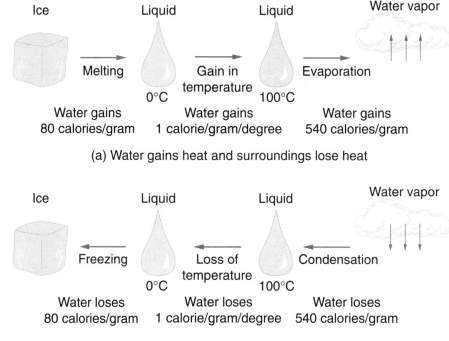

(a) Water gains heat and surroundings lose heat

(b) Water loses heat and surroundings gain heat

water evaporates, the water gains heat and the air loses heat.

Condensation is the change of water vapor to liquid water. When water vapor condenses in a cloud, it forms tiny water droplets. The water vapor loses heat and the air gains heat.

Sublimation is the change of ice directly to water vapor. On very warm sunny days after a snowstorm, some snow will change directly into water vapor. The snow gains heat and the air loses heat.

Deposition is the change of water vapor directly into ice. When the air temperature is below freezing in a cloud, the water vapor changes directly into snow crystals (ice). The water vapor loses heat and the air gains heat.

Activity

1. Place a pot of water on the stove. Heat the pot of water until you see a cloud of tiny water droplets forming over the water. Carefully observe the cloud. What type of phase change occurred to cause the water in the pot to change into water vapor above the pot?

2. The water vapor above the pot immediately cools and produces a cloud of tiny water droplets. What type of phase change occurred to cause the water vapor to change into tiny water droplets to form the cloud?

3. As the cloud rises it disappears. What type of phase change occurred to cause the water droplets to change back into water vapor in the air?

4. Describe the heat exchange when water droplets change to water vapor in the air.

SKILL EXERCISE—*Putting Data in a Chart*

*I*nformation can be put in an orderly form by creating a chart or a table. For instance, information about heat gain and heat loss between water and air can be organized in the form of a table. A table of information is usually much easier to read than text.

Design a chart listing the six water phase changes that occur in the atmosphere. For each phase change state 1) the number of calories per gram needed, 2) whether water gains or loses heat, and 3) whether the air gains or loses heat. The information for the chart can be found in the text. You may have difficulty determining the calories per gram for sublimation and deposition. Remember during sublimation and deposition the heat must be added for the liquid state also.

Interesting Facts About Water in the Atmosphere

How much water is in the atmosphere? It depends upon the temperature and the amount of water available. Under average temperature and humidity conditions, there is about 10 g of water vapor in every cubic meter of air. A room 3 meters high by 10 meters long by 10 meters wide contains 300 cubic meters of air. If there were 10 g of water vapor per cubic meter, there would be 3000 g, or 3 liters of water in the air.

How long does water stay in the atmosphere? A molecule of water stays about 9 days in the atmosphere before it returns to Earth's surface, usually in the form of rain.

What percent of Earth's water is in the atmosphere? About 0.001 percent of Earth's water is vapor in the atmosphere; that is 1 in every 100,000 liters of water in the atmosphere.

If all the water vapor in the atmosphere condensed, how much water would be added to Earth's surface? If all the water in the atmosphere condensed, about 2.5 cm of water would cover the entire Earth.

Questions

1. What are the three phases of matter?
   (1) solid, liquid, and gas
   (2) air, fire, and ice
   (3) snow, rain, and sleet
   (4) water, rock, and air

2. During the process of freezing liquid water into ice
   (1) the water gains heat energy
   (2) the water loses heat energy
   (3) the amount of heat in the water remains the same
   (4) the water loses heat energy and then gains heat energy

3. A wet shirt is put on a clothesline to dry on a sunny day. The shirt dries because water molecules
   (1) gain heat energy and evaporate
   (2) gain heat energy and condense
   (3) lose heat energy and evaporate
   (4) lose heat energy and condense

4. In which process does water absorb the greatest number of calories?
   (1) raising 1 gram of water 20°C
   (2) changing 1 gram of ice to liquid water
   (3) changing 1 gram of liquid water to water vapor
   (4) changing 1 gram of liquid water to ice

5. During the phase change of ice to liquid water the molecules
   (1) move faster and farther apart
   (2) move slower and farther apart
   (3) move faster and closer together
   (4) move slower and closer together

Thinking and Analyzing

1. Use your knowledge of science and the diagram below to answer the following questions. The diagram shows a phase change represented by letter *A*.

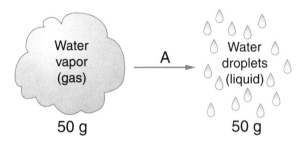

a. What is the term for the phase change represented at *A*?
b. Describe the heat exchange when water vapor changes to water droplets.
c. How much heat is exchanged during the phase change?

2. The diagram below illustrates how water is cycled in and out of the atmosphere. At location *A*, water from the lake is evaporating into the air. At location *B*, the water vapor in the air is condensing into tiny water droplets that form a cloud. At location *C*, rain is falling to Earth, refilling the lake.

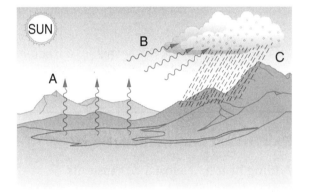

a. Describe the phase change occurring at location *A*.
b. Is the water gaining or losing heat energy during the phase change at location *A*?
c. Describe the phase change occurring at location *B*.
d. Describe the heat exchange at location *B*.

What Is Expansion and Contraction?

Objectives
Explain why substances expand and contract.

Explain how expansion and contraction affect the construction of bridges.

Terms
expansion (ik-SPAN-shuhn): an increase in the length or volume of a substance

contraction (kahn-TRAK-shuhn): a decrease in the length or volume of a substance

Expansion and Contraction

Adding heat to a substance causes its molecules to vibrate more rapidly and move farther apart, so that the substance expands. **Expansion** is an increase in the length or volume of a substance. Removing heat does the opposite. It causes the molecules to vibrate slower, move closer together, and the substance to contract. **Contraction** is a decrease in the length or volume of a substance.

Rate of Expansion

Most solids, liquids, and gases expand when heated. Do you recall from Chapter 2 that water at 0°C is an exception? A solid object's rate of expansion is very small. For example, steel expands by 12 millionths (0.000012) of its original length for each Celsius degree its temperature increases. This is a very small amount of expansion, but when building large structures like bridges and buildings, a small amount of expansion is critical. Do you know why? (See the next section.)

Liquids also expand a very small amount when heated. A cooling system in a car contains a liquid solution. When the solution heats it expands. The expanded solution would cause the car's radiator and hoses to burst if they were in a closed system. An expansion tank allows the cooling solution to be released into a reservoir without damage to the car's cooling system.

Gases, like solids and liquids, also expand a very small amount when heated. Air in contact with Earth's surface is heated primarily by conduction. As already discussed, the air expands, becomes less dense, and rises. You can often see rising air as blurry waves above a burning fire or hot pavement surface.

How Expansion and Contraction Affect Us

Solid objects, such as railroad tracks, bridges, and sidewalks expand in the summer and contract in the winter. They all have expansion and contraction joints that allow them to freely expand and contract due to temperature changes. Railroad tracks are constructed with a gap between each rail. (See Figure 4.6-1.) When the temperature is cold, the rail contracts and the gap gets wider. When the temperature is hot, the rail expands and the gap gets narrower. If the rail was heated and there was no expansion gap, the track would buckle and twist.

An engineer designing a bridge must take into consideration the expansion rates of the materials used to construct the bridge. For example, the steel beam on the Verrazano-Narrows Bridge is about 1300 meters long. An annual temperature change of 50°C causes the bridge to expand almost 1 meter in length. Without gaps called expansion joints on the bridge, the bridge would buckle. Next time you cross a bridge, look for the expansion joints in the roadway.

Although most substances expand when heated and contract when cooled, water is an exception. At most temperatures liquid water will expand when it gains heat and contract when it loses heat. However, when

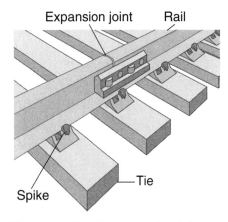

Figure 4.6-1. An expansion joint can be found between rails on a railroad track.

water is cooled down to 4°C it stops contracting and starts to expand. It continues expanding until it becomes ice at 0°C. This is why an unopened glass bottle of water or soda will crack if left outdoors during freezing weather. The force generated by the expansion of water changing to ice is so powerful that it can crack glass, rocks, concrete, and even steel.

When water freezes its volume increases and its density decreases. (See Chapter 2.) The density of ice becomes less than the density of water. Anything with a density less than that of water floats. Therefore, ice floats on top of water.

Activities

1. To do this activity you need a balloon, a small plastic (soda) bottle, and hot water. Place the opening of the balloon over the neck of the plastic bottle. Place the bottle in hot water.

a) What do you predict will happen to the balloon? Why?

b) What does this activity demonstrate?

2. To do this activity you need a small plastic (soda) bottle with a cap, a nail and hammer, clay, a drinking straw, and hot and cool water. Remove the bottle cap from the bottle. Using the nail and hammer, punch a hole through the bottle cap large enough to place the straw through it. Push the straw about half way through the hole in the bottle cap. Fill the bottle to the top with cool water. Put the bottle cap on the bottle and place the clay around the straw so air cannot pass in or out of the bottle. Place the bottle in hot water as shown in the figure below.

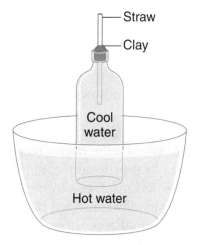

a) What do you predict will happen to the water in the bottle? Why?

b) What does this activity demonstrate?

Interesting Facts About Ice

One reason Earth is a unique planet in the solar system is because it contains water. Water is the only common substance on Earth that can be found as a liquid, solid, and gas at normal Earth temperatures. But, even more amazing is what happens to water when it changes to ice.

Like most other substances, when liquid water is cooled its molecules slow and it contracts. Its volume decreases and its density increases. However, when the temperature gets low enough, something strange happens. Near 0°C its volume suddenly increases and its density decreases. Its volume increases by about 10 percent. A decrease in density allows ice to float on water. You see this every time you place ice cubes in a glass of water or soda.

If it were not for this unusual property of water, life on Earth would be much different or not exist at all. If ice were denser than water, it would form at the bottom of lakes and rivers. Bodies of water would freeze from the bottom up, completely killing all plants and animals living in the water. Summer heating would melt only the top layer of the lake, and ice would exist year round in the bottom layers.

It is suggested by some scientists that life on Earth would not be possible if ice were denser than water. In any case, the fact that ice floats rather than sinks is a big reason why life exits on Earth.

Questions

1. Adding heat to a substance causes its molecules to
   (1) vibrate more rapidly and move farther apart
   (2) vibrate more slowly and move farther apart
   (3) vibrate more rapidly and move closer together
   (4) vibrate more slowly and move closer together

2. When heated the rate of expansion for solid, liquids, and gases is
   (1) exactly the same for all three
   (2) very small for all three
   (3) very large for all three
   (4) constantly changing for all three

3. Steel beams in a skyscraper expand and contract because
   (1) of their long length
   (2) they contain expansion and contraction joints
   (3) of a change in temperature
   (4) they are high above the ground

4. Expansion gaps are placed in bridges to allow for temperature changes that cause expansion and contraction. During what season would you expect the gaps to be widest?
   (1) summer   (3) fall
   (2) winter   (4) spring

5. When water freezes its
   (1) volume increases and its density decreases
   (2) volume decreases and its density increases
   (3) volume increases and its density increases
   (4) volume and density remain constant

Thinking and Analyzing

1. The illustration below shows a solid metal ball and a ring before and after heat is applied to the metal ball. Before heat is applied, the metal ball passes through the ring. After heat is applied, the metal ball does not pass through the ring.

a) What heat transfer process heated the metal ball?
b) Describe the physical change in the ball. Why did the physical change occur?

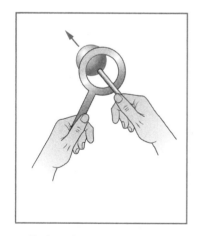
Before heating metal ball

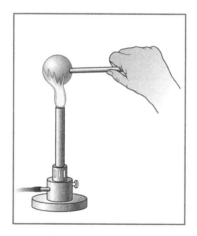
Heating metal ball

After heating metal ball

Review Questions

Term Identification

Each question below shows two terms from Chapter 4. One of the terms is defined.
(1) Choose the term that matches the definition.
(2) Describe how the two terms are different. Following each term is the section (in parenthesis) where the description or definition of the term is found.

1. *Heat (4.1) – Temperature (4.1)*
   A form of energy produced by vibrating molecules in a substance

2. *Convection (4-4) – Conduction (4-3)*
   The transfer of heat by direct contact from one object to another object

3. *Long-wave radiation (4.2) – Short-wave radiation (4.2)*
   Type of radiation transmitted from a hot heat source (sun)

4. *Expansion (4.6) – Contraction (4.6)*
   A decrease in the length or volume of a substance

5. *Troposphere (4.3) – Stratosphere (4.3)*
   The lowest layer of the atmosphere where all weather takes place

6. *Condensation (4.5) – Evaporation (4.5)*
   The process of changing water vapor to liquid water

Multiple Choice (Part 1)

Choose the response that best completes the sentence or answers the question.

1. There are four beakers on the lab table. Each beaker contains 200 mL of water. The temperature of the water in each beaker is recorded. Which beaker contains water that has the fastest vibrating molecules?

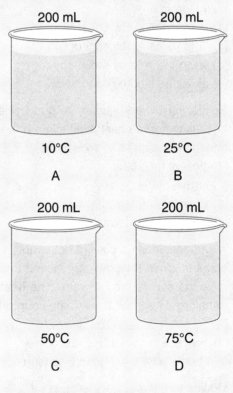

(1) Beaker *A*   (3) Beaker *C*
(2) Beaker *B*   (4) Beaker *D*

2. A thermometer is placed in a cup of hot water. The thermometer measures the
   (1) amount of heat in the water
   (2) average molecular motion of the water
   (3) amount of heat the water is losing
   (4) amount of heat the water is gaining

3. The kinetic molecular theory of matter states that matter is composed of tiny particles called molecules, and these molecules are
   (1) always stationary
   (2) traveling in a single direction
   (3) in constant vibrating motion
   (4) in motion only when they are heated

4. When cool air passes over a warm land surface there is an exchange of heat. Heat moves from the
   (1) land to the air by convection
   (2) land to the air by conduction
   (3) air to the land by radiation
   (4) air to the land by conduction

5. Solids, liquids, and gases all conduct heat, but solids are the best conductors of heat because molecules in solids are
   (1) closest together
   (2) farthest apart
   (3) vibrating the fastest
   (4) vibrating the slowest

6. A student placed a bowl of hot soup on the counter. When the bowl was moved to the table the countertop felt warm. The heat transferred from the bowl to the countertop by
   (1) convection   (3)radiation
   (2) conduction   (4)condensation

7. Liquids and gases transfer heat by
   (1) convection   (3) conduction
   (2) radiation   (4) waves

8. How is heat energy transferred within the liquid water in a pond?
   (1) compound formation
   (2) convection currents
   (3) chemical reactions
   (4) nuclear reactions

9. Radiation from the sun travels through space
   (1) in the form of waves
   (2) by direct contact of molecules
   (3) by motion through a liquid
   (4) by motion through a gas

10. Almost all heat radiation on Earth comes from
    (1) volcanoes   (3) fire
    (2) the sun   (4) inside Earth

11. A substance has a freezing point of −38°C and a boiling point of 356°C. At what temperature would this substance be in a liquid state?
    (1) −100°C   (3) 80°C
    (2) −50°C   (4) 375°C

12. Which statement describes how heat is exchanged when ice is placed in a glass of warm soda?
    (1) The ice loses heat and the soda gains heat.
    (2) The ice loses heat and the soda loses heat.
    (3) The ice gains heat and the soda gains heat.
    (4) The ice gains heat and the soda loses heat.

13. Continuously adding heat to ice will cause the ice to
(1) melt, increase temperature, and then boil
(2) melt, decrease temperature, and then boil
(3) freeze, increase temperature, and then boil
(4) freeze, decrease temperature, and then boil

14. Expansion is due to
(1) an increase in the size of molecules
(2) a decrease in the size of molecules
(3) a decrease in the distance between molecules
(4) an increase in the distance between molecules

15. The Law of Conservation of Energy states that energy is
(1) created by the sun
(2) created by Earth
(3) created by fire
(4) not created

Thinking and Analyzing (Part 2)

1. Base your answers to question 1 on the diagram below, which show four beakers containing different amounts of water at different temperatures.

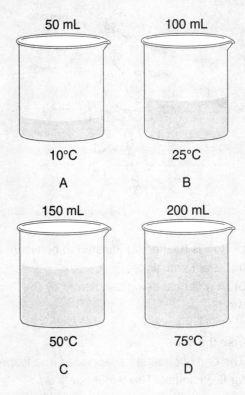

a) Which beaker contains water with the greatest average molecular motion? Explain.

b) Which beaker contains water with the least amount of heat?

c) What two factors must be considered when determining the amount of heat a substance contains?

2. Base your answers to question 2 on the diagram below, which shows a boy walking barefoot down the street on a hot summer day. The wind is blowing from left to right.

a) How is heat being transferred between the street and his feet?

b) How is heat being transferred by the wind?

c) How is heat being transferred to Earth from the sun?

3. The end of a metal bar is placed in a flame for five minutes. The temperature is measured by thermometers at four points on the bar, as shown in the diagram.

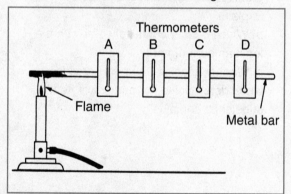

a) What method of heat transfer is illustrated in this demonstration?

b) Which thermometer would most likely record the lowest temperature? Explain.

c) What physical change occurs in the metal bar?

4. Base your answers to this question on the diagram below, which shows a room with a fire in the fireplace.

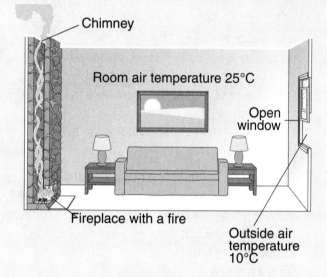

a) Describe the heat flow in the room. How would you draw an arrow to show the air movement in the room?

b) Why does heat go up the chimney?

c) What method of heat transfer does this diagram represent?

5. Base your answers to this question on the diagram below, which shows how heat radiation is trapped in a glass structure.

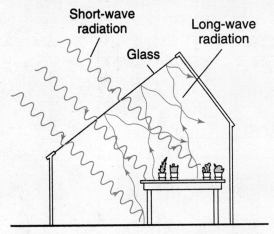

Short-wave radiation

Long-wave radiation

Glass

a) Where do the short-wave radiation and long-wave radiation originate?
b) What is this process called?
c) Describe how this process occurs in Earth's atmosphere.

6. Base your answers to this question on the diagram below, which shows a demonstration set up by a 6th grade science teacher.
a) What phase change of water is occurring in the beaker? At what temperature does this phase change occur?
b) What method of heat transfer is distributing heat in the water?
c) Predict what would happen to the water in the beaker if the teacher allowed this demonstration to continue.

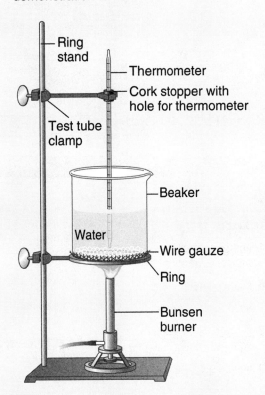

Ring stand

Thermometer

Cork stopper with hole for thermometer

Test tube clamp

Beaker

Water

Wire gauze

Ring

Bunsen burner

Chapter Puzzle (*Hint:* The words in this puzzle are terms used in the chapter.)

Across

1 energy related to molecular motion

6 change liquid to gas

7 solid form of water

9 _____change, such as liquid to gas

10 solid, liquid, and _____

14 heat transfer by direct contact

15 unit of measure for temperature

16 heat radiation

17 change solid to liquid

18 _____ molecular theory

19 heating effect caused by trapped infrared waves

Down

2 measure of average molecular motion

3 substance that flows

4 layer of atmosphere containing weather

5 heat transfer in a fluid

8 increase in length or volume

11 distance from top of one wave to top of next wave

12 transferring energy in waves

13 decrease in length or volume

14 unit of measure for heat energy

Chapter 5

Weather

Contents

Atmospheric conditions can range from very calm to very violent.

What Is This Chapter About?

Weather affects you every day. You check the weather forecast in the morning to decide how to dress for the day. Rain may affect your outdoor plans, and snowstorms may cause schools to close.

In this chapter you will learn:

1. Weather is the state of the atmosphere at a given location over a short time.

2. The sun is Earth's major source of heat. Unequal heating of Earth's surface is the primary cause of weather.

3. Some elements of the atmosphere that are used to describe weather are temperature, air pressure, wind, humidity, and precipitation.

4. The hydrosphere (Earth's water) and the atmosphere (Earth's air) constantly interact with each other.

5. Air masses are large bodies of air with similar temperatures and humidity.

6. Weather fronts are produced along the leading edge of air masses.

7. Storms are major disturbances in the atmosphere.

Career Planning: Meteorologist

Meteorologists are scientists who study the weather and climate. They predict weather changes by observing how the environment interacts with the atmosphere. Some meteorologists predict or forecast the weather on television, on the radio, or in the newspaper. Other meteorologists conduct research for the government. Some companies hire meteorologists to study air pollution, global climate changes, and how the ocean interacts with the atmosphere.

Internet Sites:

www.nws.noaa.gov Provides current weather information and links to other weather sites.

www.weather.com Connected to the Weather Channel, provides up-to-date weather information throughout the US, and general information about many weather topics.

www.accuweather.com Provides current weather information for cities and countries, with links to current weather events and topics.

5.1

What Causes Weather?

Objectives
Describe how the sun heats Earth's land, water, and air.

Explain what causes weather.

Terms
radiant (RAY-dee-uhnt) **energy:** energy, such as light, radio waves, and X-rays, that is capable of passing through space

weather: the state of the atmosphere at a given location over a short period of time

The Atmosphere

The atmosphere consists of gases that are held to Earth's surface by gravity. About 97 percent of the gases are located in the lowest 30 kilometers of the atmosphere. Although there is no boundary to the upper limit of the atmosphere, gases thin out sufficiently to suggest that the top of the atmosphere is about 150 km above the surface.

Table 5.1-1 shows the average composition of Earth's atmosphere. The

Table 5.1-1. Composition of Earth's Atmosphere

Nitrogen	78%
Oxygen	21%
Carbon Dioxide	0.03%
Other Gases	0.17%
Water Vapor (variable)	1–3%

composition of the atmosphere is relatively constant. Although some processes, such as respiration and fires, remove oxygen from the atmosphere, other processes, such as photosynthesis within green plants, add oxygen to the atmosphere. (See Lesson 7.8.)

Radiant Energy From the Sun

Earth receives almost all its energy from the sun. Light from the sun is a form of **radiant energy** that travels through space and strikes Earth. You can feel radiant energy when you stand in bright sunlight.

When radiant energy strikes Earth's surface, it can be absorbed or reflected. About 50 percent of the radiant energy that strikes Earth's surface is absorbed in the form of heat. (See Figure 5.1-1.) Surfaces differ in their ability to absorb heat. Dark

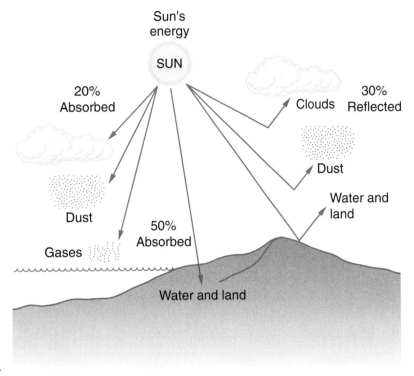

Figure 5.1-1. The distribution of radiant energy from the sun.

surfaces are better absorbers of heat than light surfaces. Land surfaces are better absorbers of heat than water surfaces. Glacial and snow-covered surfaces absorb very little heat energy. They reflect most of the energy they receive back into space.

So, how is the atmosphere heated?

- About 20 percent of the sun's radiation is absorbed directly into the atmosphere by gases, clouds, and dust particles.
- Radiation is released back into the atmosphere from Earth's land and water surfaces.
- Heat is transferred by conduction in those places where the atmosphere makes contact with the surfaces of land and water.

Unequal Heating of Earth's Surface

Weather is the state of the atmosphere at a given location over a short period of time. Weather is described by its elements—temperature, air pressure, winds, moisture content, and clouds. It is the unequal heating of Earth's surface that causes these weather elements to change.

As stated in the previous section, different types of Earth's surfaces absorb different amounts of heat. For example, buildings and streets within a city absorb more heat than the trees in a forest region. The air over a city becomes warmer than the surrounding air. Unequal air temperatures cause the air to move. Warm air rises, cool air sinks, and some air moves horizontally.

Moving air causes other changes in the weather. Rising air forms clouds and precipitation. Large amounts of rising air can produce storms such as thunderstorms, hurricanes, and tornadoes.

Activity

1. Obtain a description of the weather from the radio, television, newspaper, or Internet.

(a) What weather elements are used to describe the weather?

(b) Is the temperature given in degrees Celsius, Fahrenheit, or both?

(c) What units are used to describe the other elements?

Questions

1. What are the two major gases in the atmosphere?
   (1) carbon dioxide and oxygen
   (2) water vapor and oxygen
   (3) nitrogen and oxygen
   (4) nitrogen and carbon dioxide

2. What type of energy do we receive from the sun?
   (1) electrical   (3) radiant
   (2) nuclear   (4) mechanical

3. A concrete surface absorbs 80% of the solar radiation that strikes it. The other 20% is
   (1) reflected
   (2) absorbed by the atmosphere
   (3) absorbed by the clouds
   (4) never reaches Earth

4. Weather is caused by
   (1) changes in air moisture
   (2) changes in air temperature
   (3) changes in air pressure
   (4) unequal heating of Earth's surface

5. Which of the following four Earth's surfaces is the best absorber of heat?
   (1) water
   (2) black pavement
   (3) grassy meadow
   (4) snow

Thinking and Analyzing

1. On a warm sunny day, the air temperature is warmer above a black pavement than above a grass lawn. Explain why.

   Base your answers to questions 2 and 3 on the graph below. This graph shows the results of an experiment comparing the cooling rates of two cans of identical size and shape. These cans are painted different colors and filled with water at 100°C. The cans were allowed to cool for 60 minutes. The temperature of the water in each can was recorded every 5 minutes.

2. Calculate the rate of cooling of the water in the black can for the first 10 minutes. Use the equation below. What is the rate of cooling for the black can?

$$\text{rate of cooling} = \frac{\text{difference in temperature of water in can (°C)}}{\text{time of cooling (min)}}$$

$$= \underline{\quad\quad} \text{°C/min}$$

3. How did the color of the cans affect the cooling rate of the water during the first 10 minutes of the experiment?

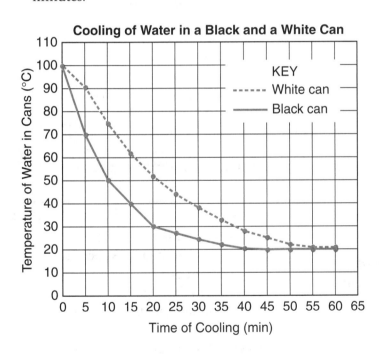

What Is Air Pressure?

Objectives

Explain the cause of air pressure.

Describe how a barometer measures air pressure.

Explain how a barometer is used to predict the weather.

Terms

air pressure: the force caused by air pushing down on an object; air pressure is exerted in all directions

density (DEHN-suh-tee)**:** how close molecules are to each other, the mass per unit volume of a substance

barometer (buh-RAHM-uht-uhr)**:** an instrument used to measure air pressure

aneroid (A-nuh-royd) **barometer:** a type of barometer that does not contain a liquid

millibar (MIH-leh-bar)**:** unit of air pressure shown on weather maps; one inch of mercury equals 33.9 millibars of air pressure

high-pressure system: cooler, denser, sinking air that usually produces clear skies and fair weather

low-pressure system: warmer, less dense, rising air that usually produces cloudy skies and rainy weather

Air Pressure

Air is a gas that is held close to the Earth by gravity. Because air has weight, the weight of the air above you produces **air pressure**. Air pressure at sea level is about 1 kg/cm^2 or 14.7lb/in^2. (See Figure 5.2-1.)

You can think of air pressure like you think of water pressure in a swimming pool. As you swim down toward the bottom of a swimming pool, you feel an increase in water pressure because the amount of water above you increases. Air pressure increases as you descend in the atmosphere. At Earth's surface, the pressure is greatest because you are at the bottom of a pool of air.

Stop and Think: What happens to air pressure as you descend below Earth's surface into a cave?

Answer: As you descend below Earth's surface into a cave, the air pressure increases even more.

The **density** of air also affects air pressure. Density refers to how close air molecules are to each other. The closer the air molecules are to each other, the denser

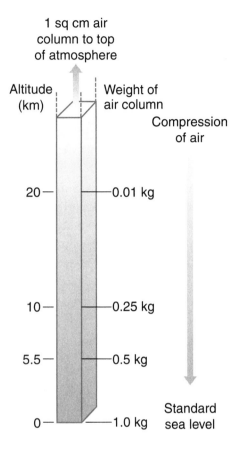

1 sq cm air
column to top
of atmosphere

Altitude
(km)

Weight of
air column

Compression
of air

20 — ┼─ 0.01 kg

10 — ┼─ 0.25 kg

5.5 — ┼─ 0.5 kg

Standard
sea level

0 — ┼─ 1.0 kg

Figure 5.2-1 The air pressure produced on a square centimeter of Earth's surface is about 1.0 kilogram.

the air. Molecules in cold air are closer together than molecules in warm air. As a result, cold air is denser than warm air. Cooler air sinks, causing an increase in air pressure. Warmer air rises, causing a decrease in air pressure.

Barometer

A **barometer** is an instrument used to measure air pressure. Early barometers consisted of glass tubes filled with liquid mercury. Today, we use other types of

barometers because we know mercury is a hazardous material that should never be handled.

How does a *mercury barometer* measure air pressure? A glass tube closed at one end is filled with mercury and placed in a dish of mercury. (See Figure 5.2-2.) The weight of the air pushing on the mercury in the dish supports the column of mercury in the tube. Standard sea-level air pressure pushes mercury up the tube to a height of 29.92 inches (760 mm). The height of the column of mercury changes as the air pressure changes. When atmospheric pressure increases, the mercury in the tube rises; when atmospheric pressure decreases, the mercury falls.

Stop and Think: What would happen to the column of mercury if you took the barometer to the top of a high mountain?

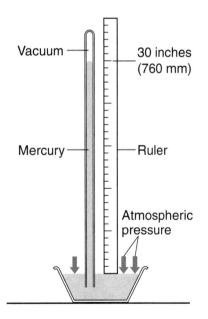

Vacuum

30 inches
(760 mm)

Mercury

Ruler

Atmospheric
pressure

Figure 5.2-2. The mercury barometer shows how atmospheric pressure pushing down into a dish of mercury causes the mercury in the tube to rise about 30 inches.

Answer: If you took the barometer to the top of a high mountain, the column of mercury would fall because the air pressure is less.

Another type of barometer is the **aneroid barometer.** An aneroid barometer does not contain a liquid. It uses a sealed air chamber with a flexible cover. Before the chamber is sealed, a small amount of air is removed from it. This allows the chamber to expand and contract with small changes in atmospheric pressure. The flexible cover is pushed down when air pressure increases, and pushed up when air pressure decreases.

A pointer is attached to the flexible cover. When air pressure changes, the pointer moves along a scale. The scale indicates the air pressure in inches of mercury, or millibars. (See the Activity and figure later in this lesson.)

On weather maps, meteorologists use units called **millibars** (mb) to show air pressure. One inch of mercury equals about 33.9 millibars of air pressure. Standard air pressure in millibars is 1013.2 mb. Air pressure readings on weather maps provide information about changing weather. *Isobars* are lines on weather maps that

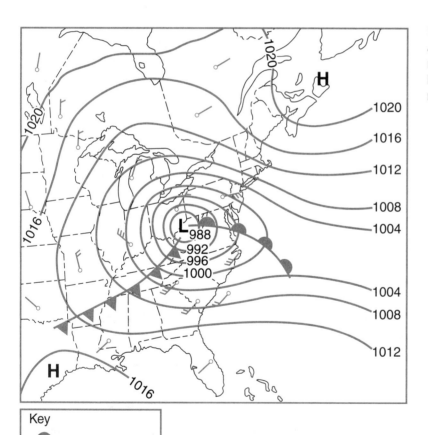

Figure 5.2-3. Lines of equal air pressure on a weather map are called isobars. Isobars are labeled in millibars of air pressure.

Key
Warm front
Cold front

connect points of equal air pressure and are usually measured in millibars. Isobars help meteorologists locate high- and low-air-pressure systems. (See Figure 5.2-3.)

Air Pressure and Weather

As we learned in the last lesson, the sun heats Earth's surface at different rates. Land heats faster than water, and dark-colored surfaces heat faster than light-colored surfaces. This unequal heating causes differences in air temperature and differences in air density. Differences in air density cause air to rise and fall. Cold, dense air sinks, producing a **high-pressure system** (**H** on a weather map). Warm, less dense air rises, producing a **low-pressure system** (**L** on a weather map).

Sinking, cool air in a high-pressure system causes winds to circulate around it in a clockwise direction (in the northern hemisphere). High-pressure systems tend to produce clear skies and fair weather. Rising, warm air in a low-pressure system causes winds to circulate in a counterclockwise direction (in the northern hemisphere). Low-pressure systems tend to produce clouds and rainy weather. (See Figure 5.2-4.)

Stop and Think: How can barometers help predict the weather?

Answer: A rising barometer usually indicates the approach of a high-pressure system and fair weather. A falling barometer usually indicates the approach of a low-pressure system and the possibility of clouds, wind, and rain.

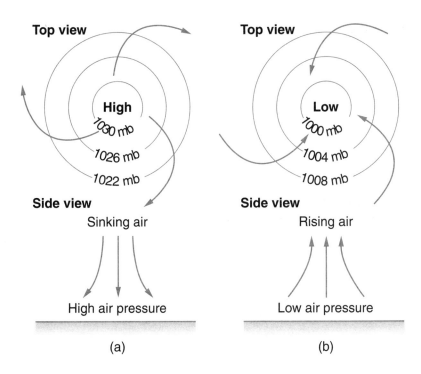

Figure 5.2-4. Movements of air currents in (a) a high-pressure system; (b) a low-pressure system (mb is the abbreviation for millibars).

Interesting Facts About Air Pressure

Air pressure can produce a tremendous force. For example, a Boeing 747 that weighs 360,000 kg is lifted by air pressure when it flies. The design of the wing and the speed of the plane allow a tremendous airflow over the wing. Low air pressure is produced above the wing, and the higher air pressure under the wing produces enough force to lift the plane.

Activity

A mercury barometer is awkward to carry from place to place. Today, most people use aneroid (not liquid) barometers to measure air pressure. The aneroid barometer uses a flexible chamber from which a small amount of air has been removed. This allows the chamber to expand and contract when it encounters slight differences in air pressure. As the outside air pressure increases, the chamber lid pushes down. The dial moves to indicate the change in air pressure on the attached scale. The figure shows how to make a simple aneroid barometer. To build this barometer, you need a 3–4-cm wide-mouth bottle, a balloon, a rubber band, glue, and one-half of a paper straw to

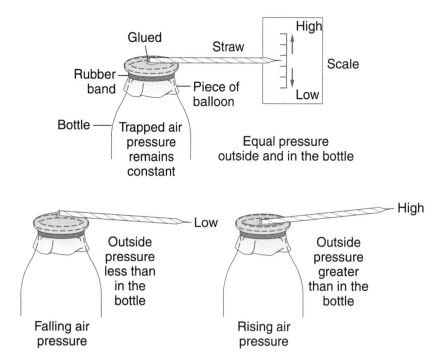

act as a pointer. It is best, but not necessary, to make the barometer on a day when the air pressure is about 30.00 inches of mercury. Attach the balloon patch over the bottle mouth with the rubber band. Glue the end of the straw to the center of the balloon patch. Pinch the end of the straw (glue if necessary) and place it next to a scale (see figure).

Observe the barometer for several days.

1. Describe what happens to the end of the straw from day to day. What is happening to the air pressure outside the bottle when the pointer end of the straw rises? What is happening to the air pressure outside the bottle when the pointer end of the straw falls?

2. What happens if you place the barometer in the sunlight? Why?

3. Do you think your barometer is accurate? Why?

SKILL EXERCISE—*Plotting a Graph*

*R*eproduce the graph below on a sheet of graph paper. On your graph, plot the data from the Altitude and Air Pressure table. Neatly and accurately, connect the points on the graph to produce a curved line.

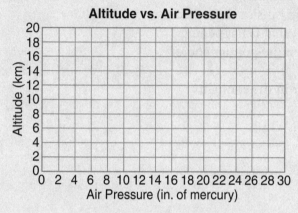

Altitude vs. Air Pressure

Altitude and Air Pressure

Altitude (km)	Air Pressure (in. of mercury)
20	1.6
18	2.2
16	3.1
14	4.2
12	5.7
10	7.8
8	10.5
6	13.9
4	18.2
2	25.3
0	29.9

Answer the following questions after you complete the graph:

1. a. What is the air pressure at 4 km?
 b. What is the air pressure at 9 km?

2. At what elevation is the air pressure 2.5 in. of mercury?

3. Describe the relationship between altitude and air pressure.

4. What happens to the air pressure as you ride an elevator to the top floor in a tall building?

Questions

1. The source of energy that sets Earth's atmosphere in motion and causes weather changes is
   (1) volcanism
   (2) gravity
   (3) the ocean
   (4) the sun

2. Compared to cool air, warm air is
   (1) more dense and rises
   (2) less dense and rises
   (3) the same density
   (4) less dense and sinks

3. In a mercury barometer the weight of the air is indicated by
   (1) the height of the mercury column in the glass tube
   (2) the amount of mercury in the dish
   (3) the length of the glass tube
   (4) the height of the barometer above sea level

4. In the figure below, the air pressure would be greatest at point
   (1) Point A   (3) Point C
   (2) Point B   (4) Point D

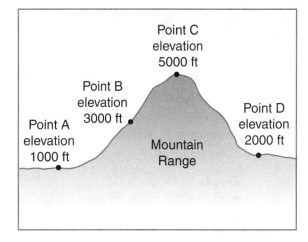

5. A rising barometer usually indicates
   (1) an approaching high-pressure system and stormy weather
   (2) an approaching high-pressure system and fair weather
   (3) an approaching low-pressure system and stormy weather
   (4) an approaching low-pressure system and fair weather

Thinking and Analyzing

1. Mr. Chang wanted to determine the change in air pressure as he drove his car up Pike's Peak, a high mountain in Colorado. What weather instrument should he take to measure the change in air pressure? Predict the change he observed.

2. The density of mercury is 13.6 times greater than the density of water. Early barometers used water instead of mercury. Determine the height of a water column if the air pressure is 29.92 inches of mercury. Describe the problem that occurred when water was used instead of mercury.

How Do Winds Form?

Objectives

Explain how winds form.

Compare land and sea breezes.

Describe the global wind pattern.

Terms

wind: the horizontal movement of air over Earth's surface

sea breeze: a cool wind that blows onshore in the daytime

land breeze: a warm wind that blows offshore at night

prevailing westerlies (pree-VALE-eeng WES-tur-lees): a wind belt in the lower atmosphere where winds blow from the southwest, generally between 30°–60°N latitude

wind chill index: the temperature we perceive due to the effect of wind

---

Formation of Winds

Wind is the horizontal movement of air over Earth's surface. Air can also rise and fall. Rising air is called an *updraft*, and falling air is called a *downdraft*.

Unequal heating of Earth's surface forms low-pressure and high-pressure areas. Warm air, because it is less dense than the surrounding air, produces an updraft and a low-pressure area. Cool air, because it is denser than the surrounding air, produces a downdraft and a high-pressure area. Air flows from high-pressure areas to low-pressure areas. Surface winds are produced by air moving from high-pressure areas to low-pressure areas. (See Figure 5.3-1.) The greater the differences in air pressure

between high-pressure and low-pressure areas, the greater the wind speed.

Sea and Land Breezes

Sea breezes and land breezes commonly form along Long Island's beaches on warm, cloudless summer days. A difference in temperature between the air over the land and the air over the water produces a difference in air pressure. This difference causes a convection current to form. (See Lesson 4.4.)

A **sea breeze** is a cool wind that blows onshore in the daytime. It forms because land absorbs the sun's heat better than water. The air over land becomes warmer than the air over water. The warmer air rises

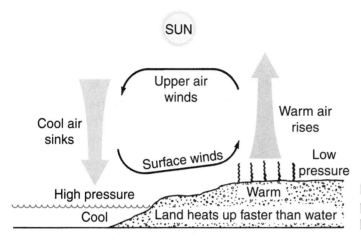

Figure 5.3-1. Surface winds are produced by air moving from a high-pressure area to a low-pressure area.

and produces a low-pressure area. The sinking air over the water produces a high-pressure area. An onshore sea breeze forms because the air moves from the high-pressure area to the low-pressure area. By mid-day, a well-formed convection current is produced. [See Figure 5.3-2(a).]

A **land breeze** is a wind that blows offshore at night. It forms because land cools more rapidly than water. The warmer air over the water rises, producing a low-

pressure area. The sinking air over the land blows offshore, replacing the rising air. Land breezes usually do not form as well as sea breezes, because the temperature difference between the land and water at night is not always great enough to produce a strong convection current. Figure 5.3-2(b) illustrates the complete circulation of air forming a land breeze.

Perhaps, on a summer day when you are at the beach, you will recognize the

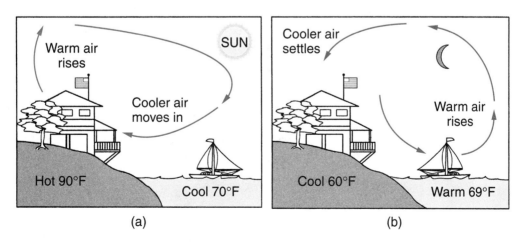

Figure 5.3-2. (a) A sea breeze forms in the daytime; (b) a land breeze forms at night.

formation of a sea breeze. In the late morning, after the sun has had time to heat the land and cause the warm air to rise, you will feel a cool breeze blowing from the ocean.

Global Winds

We can see winds blowing the leaves on trees or the hats off people, but a larger wind system exists on Earth that is more difficult to observe. This wind system occurs over a global area. These winds are caused by temperature differences between the poles and the equator.

Warm air exists near the equator and cool air exists near the poles. This causes the air near the equator to be less dense and rise. The air near the poles is dense air and it sinks. On a stationary planet, this would produce global winds that blow from the poles toward the equator. [See Figure 5.3-3(a).]

However, Earth's rotation causes some complications to this simplified wind pattern. First, moving winds in the Northern Hemisphere curve to the right. Second, separate areas of high and low pressure form between the equator and the poles and cause the single convection current to be split into three smaller convection currents in each hemisphere. [See Figure 5.3-3(b).]

Most of the United States is located in the prevailing westerlies wind belt. The **prevailing westerlies** blow generally from the southwest between 30°–60°N latitude. They cause air masses, fronts, and storms to move across the United States from west to east.

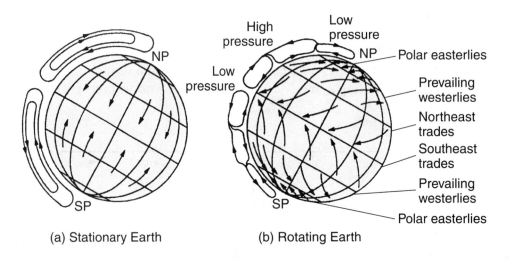

(a) Stationary Earth   (b) Rotating Earth

Figure 5.3-3.   (a) Shows an imaginary global wind pattern on Earth if Earth were stationary; (b) shows the real global wind patterns formed on Earth as it rotates.

Activities

1. It is possible to estimate wind speed using common sea and land conditions. The Beaufort Scale of Wind Force was designed in 1805 by Sir Francis Beaufort. It is still used today by mariners and pilots. Review the first seven categories in the Beaufort Scale below and try to determine the wind speed outside your home or school. Check your accuracy with the latest weather report.

2. There are many special winds that affect different areas around the world. Many of these special winds have names. Using the Internet or other reference resources, determine the location for each of the following winds.

   a. Blue Norther
   b. Chinook
   c. Foehn
   d. Monsoon
   e. Mistral
   f. Bora
   g. Zonda
   h. Tourments
   i. Buran
   j. Datoo

Beaufort Scale of Wind Force (7 of 12 categories)

No.	Wind	Speed (mph)	Sea/Land Effect
0	Calm	0–1	Sea is like a mirror. Smoke rises vertically.
1	Light Air	1–3	Ripples appear like scales on water surface. Land smoke drifts.
2	Light Breeze	4–7	Ripples consistent on water surface. Rustling leaves. Feel wind on face.
3	Gentle Breeze	8–12	Large wavelets with some foam and whitecaps. Leaves and twigs in constant motion. Flag extended.
4	Moderate Breeze	3–18	Small waves with whitecaps. Small branches move. Dust and paper are blown.
5	Fresh Breeze	19–24	Moderate waves with many whitecaps. Small trees sway.
6	Strong Breeze	25–31	Larger waves form with spray. Large branches move. Difficult to use umbrella.

SKILL EXERCISE—*Reading a Chart*

Why does it feel much colder when the wind blows on a cold day? The air temperature alone does not tell you how cold it feels outside. Wind causes the body to lose heat, and when the body loses heat it feels even colder. The effect wind has on how we perceive cold is called the **wind chill index**. It is a better indicator than temperature on how you should dress for the outdoors. The wind chill index is determined using the chart below. If the temperature is 20°F and the wind is blowing at 25 mph, it feels like it is 3°F.

Wind Chill Chart

Temperature (°F)

		40	35	30	25	20	15	10	5	0	−5
	5	36	31	25	19	13	7	1	−5	−11	−16
	10	34	27	21	15	9	3	−4	−10	−16	−22
Wind	15	32	25	19	13	6	0	−7	−13	−19	−26
Speed	20	30	24	17	11	4	−2	−9	−15	−22	−29
(mph)	25	29	23	16	9	3	−4	−11	−17	−24	−31
	30	28	22	15	8	1	−5	−12	−19	−26	−33
	35	28	21	14	7	0	−7	−14	−21	−27	−34
	40	27	20	13	6	−1	−8	−15	−22	−29	−36

1. Determine the wind chill index for each of the following situations:

(a) a temperature of 35°F and a wind speed of 15 mph

(b) a temperature of 10°F and a wind speed of 35 mph

(c) a temperature of 28°F and a wind speed of 10 mph

(d) a temperature of 22°F and a wind speed of 22 mph

2. Describe the relationship between wind speed and the wind chill index.

Interesting Facts About High Wind Speeds

The highest surface wind speed ever recorded was 231 mph. It was recorded on Mount Washington, New Hampshire, on April 12, 1934. Mount Washington (6,288 ft) is the highest point in New Hampshire.

It is believed that wind speeds in some tornadoes approach 300 mph. However, it is difficult to measure the wind speed in a tornado because weather instruments cannot withstand the force of such high wind speeds. Hurricanes seldom exceed 200 mph, and wind gusts within thunderstorms are usually less than 100 mph.

An upper air wind, the jet stream, flows like a wind tube across the United States. It is a curving wind current that flows from west to east at a speed of about 50 mph in the summer and 150 mph in the winter. Sometimes the speed of the jet stream reaches 300 mph. The jet stream is a few hundred miles wide and about six miles above Earth's surface. Airplanes that fly in the jet stream from west to east shorten their flight time. However, when an airplane flies from east to west in the jet stream, it increases the flight time. Why?

Questions

1. Differences in air pressure on Earth's surface are produced by
   (1) differences in moisture content
   (2) differences in air composition
   (3) unequal heating
   (4) clouds

2. Which global winds affect the weather in New York City?
   (1) polar easterlies
   (2) prevailing westerlies
   (3) northeast trade winds
   (4) southwest trade winds

3. Winds blow from
   (1) high-pressure areas to low-pressure areas
   (2) low-pressure areas to high-pressure areas
   (3) high-pressure areas to high-pressure areas
   (4) low-pressure areas to low-pressure areas

4. A sea breeze is most likely to form on a
   (1) warm, cloudless, summer day
   (2) cool, cloudy, summer day
   (3) cool, cloudy, winter day
   (4) warm, cloudy, winter day

5. Weather systems move across the United Stated from
   (1) north to south
   (2) south to north
   (3) east to west
   (4) west to east

Thinking and Analyzing

Explain why it is necessary to look at the weather west of New York State to forecast tomorrow's weather.

5.4

What Is Relative Humidity?

Objectives

Explain what relative humidity is.

Describe how relative humidity is determined using the relative humidity chart.

Terms

water vapor: the gaseous form of water in the air

humidity (hyoo-MIHD-ih-tee): the amount of water vapor in the air

evaporation (ee-vap-uh-RAY-shuhn): a phase change of a liquid into a gas

maximum humidity: the greatest amount of water vapor the air can hold based upon the temperature of the air

relative humidity: the ratio between the actual amount of water vapor in the air, and the maximum amount of water vapor the air can hold at a specific temperature

psychrometer (sy-KRAHM-uh-tuhr): an instrument that uses a wet-bulb and a dry-bulb thermometer to measure relative humidity

Humidity

Water vapor is the gaseous form of water. **Humidity** refers to the amount of water vapor in the air. Hose down the sidewalk on a sunny day, and within a short time the sidewalk will be dry. Where does the water on the wet sidewalk go? Liquid water changes into water vapor by the process of **evaporation**, and rises into the air. The atmosphere receives water vapor from many sources. Oceans, rivers, lakes, moist soil, plants, and animals all add water vapor to the atmosphere.

Warm air can absorb or hold more water vapor than cold air. Therefore, the temperature of the air determines how much water vapor it can hold. The greatest amount of water vapor air can hold at a specific temperature is called its **maximum humidity**. A cubic meter of air at 18°C can hold about 20 grams of water, and air at 25°C can hold bout 29 grams of water.

Stop and Think: What can you predict about the maximum humidity of air at 15°C?

Answer: Air at 15°C holds less than 20 grams of water vapor per cubic meter.

The actual amount of water vapor in the air is usually less than the maximum amount the air can hold. Under certain conditions, however, air is filled with water vapor and can hold no more. This is called *saturated* air. When this happens, the actual amount of water vapor in the air is equal to the maximum amount of water vapor it can hold at that temperature. The air is saturated when the weather conditions are foggy and rainy.

Relative Humidity

Relative humidity is the ratio of the actual amount of water vapor in the air to the maximum amount of water vapor the air can hold at that temperature. If you know the actual humidity and maximum humidity, you can determine the relative humidity using the formula:

$$\text{Relative Humidity} = \frac{\text{Actual Humidity}}{\text{Maximum Humidity}}$$

A cubic meter of air at 18°C can hold about 20 grams of water vapor (maximum humidity). Suppose there are only 10 grams of water vapor in the air (actual humidity). The relative humidity is determined by dividing the actual humidity by the maximum humidity and multiplying the answer by 100 to change it to a percent.

$$\text{Relative Humidity} = \frac{\text{Actual Humidity}}{\text{Maximum Humidity}} = \frac{10g}{20g}$$

$$= \frac{1}{2} \times 100 = 50\%$$

Stop and Think: What is the relative humidity if only 5 grams of water vapor are in the air at 18°C?

Answer: The relative humidity is 25% (5/20 = .25 × 100 = 25%). When air is saturated the actual humidity and maximum humidity are the same, and the relative humidity is 100 percent. Under conditions of 100% humidity, it is foggy and usually raining (but not always).

Determining Relative Humidity

Relative humidity is the term most often used to express humidity. We measure relative humidity with an instrument called a **psychrometer**. One type of psychrometer consists of two liquid thermometers. One thermometer is called a dry-bulb thermometer and measures the present air temperature. The other thermometer is called a wet-bulb thermometer. The bulb of the wet-bulb thermometer is wrapped with a wet cloth and measures the cooling effect evaporation has on the temperature. (See Figure 5.4-1.)

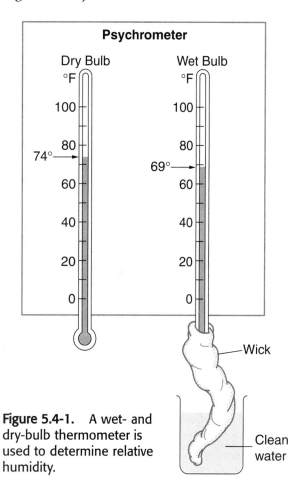

Figure 5.4-1. A wet- and dry-bulb thermometer is used to determine relative humidity.

In Chapter 4, we learned that evaporation is a cooling process. When water evaporates it gains heat and its surroundings lose heat. When air contains so much water vapor that it has nearly reached its maximum humidity, the evaporation process slows. At this point, the air is almost full of water vapor and cannot hold much more. When this condition exists, the temperature of the wet-bulb thermometer in a psychrometer decreases only slightly because very little evaporation is occurring. When the air is dry, the evaporation process increases, because the air can hold much more water vapor than it is already holding. When this condition exits, the wet-bulb temperature shows a large decrease because a great deal of evaporation is occurring.

Knowing the wet- and dry-bulb temperatures and using Table 5.4-1, you can determine the relative humidity by following steps 1–5.

1. Obtain the wet- and dry-bulb temperature readings in degrees Fahrenheit (°F).
2. Subtract the wet-bulb temperature from the dry-bulb temperature to obtain the difference in temperature.
3. Using Table 5.4-1, locate the dry-bulb temperature on the left side of the table.
4. Locate the difference in readings at the top of the table.
5. You can locate the relative humidity where the two items meet on the chart. Relative humidity is expressed as a percent.

What is the relative humidity if the dry-bulb is 74°F and the wet-bulb is 69°F? Subtract 69° from 74° and you obtain a difference of 5°. Find the dry-bulb reading on the left side of the table and move to the right. Locate the difference of 5° on the top of the table and move down. The two readings cross at 78. The relative humidity is 78 percent.

Table 5.4-1. Determining Relative Humidity (%)

Dry Bulb (°F)	Difference (Dry Bulb – Wet Bulb) (°F)							
	1°	2°	3°	4°	5°	6°	7°	8°
70°	95	90	86	81	77	72	68	64
72°	95	91	86	82	78	73	69	65
74°	95	91	86	82	78	74	70	66
76°	96	91	87	83	78	74	70	67
78°	96	91	87	83	79	75	71	67
80°	96	91	87	83	79	76	72	68
82°	96	91	87	83	79	76	72	69
84°	96	92	88	84	80	77	73	70
86°	96	92	88	84	80	77	73	70
88°	96	92	88	85	81	78	74	71
90°	96	92	88	85	81	78	74	71

Activities

Practice using Table 5.4-1 to determine relative humidity. Determine the relative humidity for each of these problems:

a. What is the relative humidity if the dry-bulb is 84°F and the wet-bulb is 82°F?

b. What is the relative humidity if the dry-bulb is 88°F and the wet-bulb is 80°F?

c. What is the relative humidity if the dry-bulb is 74°F and the wet-bulb is 74°F?

Questions

1. What determines the maximum amount of water vapor the air can hold?
   (1) the relative humidity
   (2) the actual humidity
   (3) the air temperature
   (4) the air pressure

2. Which ratio defines relative humidity?
   (1) wet-bulb reading / dry-bulb reading
   (2) actual humidity / maximum humidity
   (3) maximum humidity / actual humidity
   (4) actual humidity / wet-bulb reading

3. To use the relative humidity table you need to know the dry-bulb reading and
   (1) the total of the dry-bulb and wet-bulb readings

(2) the difference between the dry-bulb and wet-bulb readings
(3) the air pressure
(4) the maximum humidity

4. What is the relative humidity if the maximum amount of water vapor the air can hold is 35 grams of water, and there are 35 grams of water in the air?
   (1) 35 percent   (3) 70 percent
   (2) 0 percent   (4) 100 percent

5. Under what weather conditions is the relative humidity most likely 100 percent?
   (1) fog or rain
   (2) sunny and clear
   (3) sunny and windy
   (4) cloudy and no rain

Thinking and Analyzing

1. The empty glass in the diagram represents the water vapor capacity of air at 25°C. Draw diagrams to show what the glass would look like if the air had a relative humidity of 100%; if it had a relative humidity of 75%; and if it had a relative humidity of 25%.

2. How would the glass change if it represented air at 40°C?

Empty glass at 25°C. The size of the glass represents how much water air can hold at 25°C.

5.5   What Is Precipitation?

Objectives

Describe how clouds form.

Describe how precipitation is formed.

Explain how different types of precipitation form.

Terms

condensation (kahn-duhn-SAY-shuhn) **level:** the altitude where the temperature is low enough to cause water vapor to condense into tiny water droplets and clouds form

condensation (kahn-duhn-SAY-shuhn): a phase change of a gas into a liquid

stratus (STRAYT-uhs): flat, layered clouds

cumulus (KYOO-myuh-luhs): puffy clouds

cumulonimbus (KYOO-myuh-low-NIM-buhs): a thunderhead; a cloud associated with a thunderstorm

precipitation (pree-sihp-uh-TAY-shuhn): liquid or solid water that falls to Earth's surface from the atmosphere; rain, snow, sleet, and hail are examples of precipitation

sleet: a form of precipitation produced when rain freezes as it falls to Earth

hail: a form of precipitation formed by layer upon layer of ice; it is commonly round or pea size, but can sometimes be very large and irregularly shaped

flood: the overflowing of water over land that is not usually under water

drought (DROWT): a lower-than-normal amount of precipitation for extended periods of time

---

Cloud Formation

We know that rising air expands and cools. What causes air to rise? Air can rise 1) along a boundary between cold and warm air, 2) by winds moving up a mountain slope, and 3) because of the unequal heating of Earth's surface. If air rises high enough it will reach the **condensation level**. The condensation level is the height in the atmosphere where the temperature is low enough to cause water vapor to **condense** into tiny water droplets and form clouds.

Clouds exist in many forms. When air rises at a low angle, flat-layered clouds called **stratus** clouds form. Vertical updrafts form

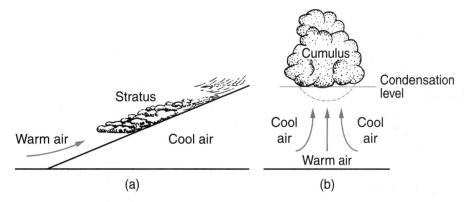

Figure 5.5-1. (a) Stratus clouds form when warm air rises at a low angle; (b) cumulus clouds form when warm air rises vertically.

puffy **cumulus** clouds. Very strong updrafts can change cumulus clouds into **cumulonimbus** clouds, or thunderheads. Cumulonimbus clouds produce thunderstorms. (See Figure 5.5-1.)

Precipitation

Water that falls out of the atmosphere is called **precipitation**. Rain is a liquid form of precipitation. Snow, sleet, and hail are solid forms of precipitation. Temperature plays a major role in the type of precipitation produced.

When warm air rises it expands, and its temperature decreases. As air cools, its ability to hold water vapor decreases. Its relative humidity reaches 100%, and some of the water vapor condenses into tiny water droplets. Clouds form from a very large number of tiny water droplets. The collision of tiny water droplets produce larger drops of water. When they become heavy enough, gravity pulls the water drops to the ground in the form of precipitation. (See Figure 5.5-2.)

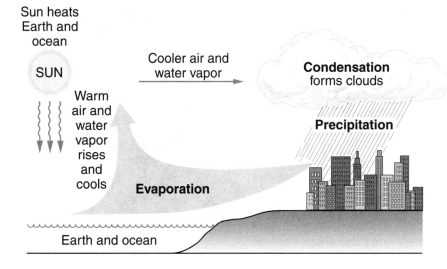

Figure 5.5-2. The path of water from ground to cloud and back down to ground.

Rain, Sleet, Snow, and Hail

Rain is precipitation in the form of liquid water. Sometimes falling rain passes through a freezing layer of air. The raindrops, which have a round shape and not a teardrop shape as is commonly thought, freeze into small irregularly shaped chunks of ice called **sleet**. Sleet is frozen rain. Sleet usually forms in the winter.

If the temperature of the air in a cloud is below freezing,, water vapor may change directly into snow crystals (deposition). When the crystals become large enough, they fall to the ground. Snowflakes may be produced if the crystals are wet, bump into other crystals, and stick together. During the winter months, clouds often contain snow crystals that melt as they fall to the ground. This produces rain instead of snow.

Thunderstorms frequently contain very strong updrafts and freezing layers of air within a cumulonimbus cloud. When raindrops fall through freezing air they change into ice. Strong updrafts may lift the frozen precipitation so more water droplets attach to it. As it falls again, it freezes again. The process can repeat many times until the ice becomes too heavy for the updraft to lift. Eventually, the ice falls to the ground as **hail**, sometimes called hailstones.

Hail is commonly round and pea sized. However, it can also be irregularly shaped and as large as a golf ball, baseball, or even a grapefruit. If you cut a large hailstone open, you can see the layers of ice that formed each time it went through the freezing layer of air. Large hail can damage crops, cars, and windows. Hail usually forms in the summer during a thunderstorm. (See Figure 5.5-3.)

Floods and Droughts

Flooding is an overflowing of water over land that is not usually under water. Heavy rainfall and rapidly melting snow and ice are the most common causes of flooding. Streams and rivers carry water from the land to the sea. If the amount of water is greater than these waterways can carry, the water overflows its banks onto the surrounding

Figure 5.5-3.   Some of the different sizes and shapes of hail.

land. Flooding can also be caused by severe storms that cause the ocean to rise onto low-lying coastal land.

Droughts are caused by a lower-than-normal amount of precipitation for extended periods of time. Other weather conditions that contribute to droughts are high temperature, low humidity, and winds. Each of these conditions causes the rapid evaporation of water from Earth's surface.

Floods and droughts are normal events that occur periodically. Flooding causes devastation by the loss of life and crops, and the destruction of buildings. Human concerns regarding drought conditions include a decrease in the freshwater supply, a lack of water for crop irrigation, and the high risk of forest fires. Floods are also beneficial. They add a layer of fertile plant nutrients to the soil. This makes flood areas excellent places to grow crops.

Activities

1. Take an absolutely dry drinking glass and place a few ice cubes in it. Then carefully place water inside the glass, making sure no water falls on the outside of the glass. Put the glass down for a minute. Touch the outside of the glass. It is wet. How did the moisture get on the outside of the glass? This activity works best on a warm humid day. Why?

2. Catch and observe snowflakes. You need a magnifying glass and some dark-colored fabric. Place the fabric in a place outdoors that will allow it to become as cold as the outdoor temperature, without becoming wet. Expose the fabric to the falling snow for a few seconds and allow a few snowflakes to fall on it. While still outdoors, observe the snowflakes with a magnifying glass. Snowflakes are a bunch of snow crystals that are stuck together. Can you see the individual snow crystals that make the snowflake? Can you see the six sides of the snow crystals? (See the photo of some shapes of snowflakes in Figure 5.5-4)

3. Observe a cumulus cloud for about 10 minutes. See if you can observe the motion of the air in the cumulus cloud. Some of the air is rising and some is falling. Try to determine if the rising air is producing the cloud or causing the cloud to disappear.

Figure 5.5-4. Snow crystals contain a variety of shapes, but they all contain six sides.

Interesting Facts About Precipitation

Blizzards are violent winter storms that are characterized by 1) very low or rapidly falling temperatures, 2) wind speeds above 35 mph, and 3) blowing snow. The snow in a blizzard might be falling snow, or snow that has been swept up from the ground by the wind. Basically, blowing snow reduces visibility and makes it difficult to see

New York City Blizzard of 1888

Most blizzards in the United States occur in the northern Great Plains and the upper Mississippi Valley. But in 1888, a major blizzard struck New York City.

Over a dozen major cities, including New York City, had their transportation and communication systems come to a halt. In southeastern New York, many areas received more than 40 inches of snow. Winds near 50 mph produced snowdrifts higher than 30 feet. In fact, in Middletown, New York, just 80 miles north of New York City, three-story houses were covered with snow. People were forced to build tunnels through the snow to get around. More than 200 people died in the New York City area. Some froze to death and others died in storm-related accidents. The Blizzard of '88 is considered one of the worst blizzards ever in the United States.

Figure 5.5-5.   The Blizzard of 1888.

Questions

1. Rising air produces a cloud when it reaches the
   (1) evaporation level
   (2) condensation level
   (3) precipitation level
   (4) 0°F level

2. Most water vapor enters the atmosphere by
   (1) condensation (3) precipitation
   (2) evaporation   (4) freezing

3. Precipitation can form as
   (1) a gas or a solid
   (2) a gas, a solid, or a liquid
   (3) a gas or a liquid
   (4) a liquid or a solid

4. Condensation of water vapor forms clouds because
   (1) cold air cannot hold as much moisture as warm air
   (2) warm air cannot hold as much moisture as cold air
   (3) cold air cannot hold any moisture
   (4) warm air cannot hold any moisture

5. If the air temperature in a cloud is above 32°F, the cloud probably consists of tiny
   (1) snow crystals
   (2) chunks of hail
   (3) chunks of sleet
   (4) water droplets

6. Droughts cause a high risk of
   (1) flooding   (3) forest fires
   (2) hurricanes   (4) earthquakes

Thinking and Analyzing

1. In January, as Mrs. Jackson was giving her science lesson, she heard what sounded like pieces of gravel hitting the classroom window. Mrs. Jackson asked her class what was causing the sound, and Shelly answered, "hail." Was Shelly correct? Explain.

2. Did you ever notice that when you talk outside on a cold day, your breath forms a tiny cloud? When this happens you are actually witnessing the process of condensation. Explain.

5.6

How Do the Hydrosphere and Atmosphere Interact?

Objectives

Explain how the hydrosphere and atmosphere interact.

Describe the water cycle.

Terms

lithosphere (LIHTH-oh-sfeer): the solid rock part of Earth

hydrosphere (HY-droh-sfeer): the surface and subsurface water on Earth

water cycle: movement of water back and forth between Earth's surface and the atmosphere by means of evaporation, condensation, and precipitation

ocean current: flowing rivers of water in the ocean

Lithosphere, Hydrosphere, and Atmosphere

Planet Earth consists of three different spheres. The three spheres are the rock sphere, or **lithosphere**; the water sphere, or **hydrosphere**; and the gaseous sphere, or *atmosphere*. (See Figure 5.6-1.)

The lithosphere consists of all the rock material in the outer layer of the Earth. The lithosphere includes the continents and the ocean floor.

The hydrosphere consists of all the water on or near Earth's surface, including water that is underground. About 75 percent of the crust is covered with oceans, lakes, rivers, and streams. By far, the greatest amount of water is in the oceans. The oceans contain 97 percent of Earth's water.

Surrounding Earth is a sphere of gases called the atmosphere. All weather takes place in the lowest 18 kilometers of the atmosphere, the *troposphere*. Great amounts of energy are exchanged in the area where the atmosphere meets the hydrosphere.

Heat Transfer Between the Hydrosphere and Atmosphere

In Chapter 4, we discussed how Earth transfers heat into the atmosphere by radiation and conduction. Heat energy is also transferred from the hydrosphere to the atmosphere by evaporation and condensation. The transfer is part of a larger process that moves water into and out of the atmosphere. This process is called the **water cycle**. The water cycle moves water back and forth between Earth's surface and the atmosphere by means of evaporation, condensation, and precipitation. (See Figure

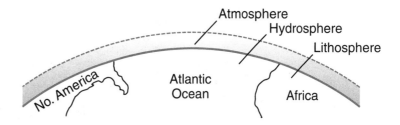

Figure 5.6-1. Earth contains a hydrosphere, lithosphere, and atmosphere.

5.6-2.) The water cycle is powered by energy from the sun.

Water absorbs heat from its surroundings when it evaporates. Air containing water vapor may eventually be heated, causing it to rise, expand, and cool. When air is cooled enough, water vapor in the air condenses into tiny water droplets and forms a cloud. During the process of condensation, the water vapor releases heat into the atmosphere. (See Figure 5.6-2.) Rain may fall from the cloud. If it reaches Earth's surface it might gain heat and evaporate again. The evaporation and condensation processes are constantly moving heat from the hydrosphere into the atmosphere.

The water cycle also refills our freshwater supply on Earth. Evaporation causes the impurities in water to remain behind, and only freshwater continues through the cycle. For example, when salty ocean water evaporates, the salt is left in the ocean and freshwater becomes water vapor. Eventually, after the condensation and precipitation processes occur, the freshwater falls to Earth and is ready for human consumption.

Waves and Currents

The boundary between the atmosphere and hydrosphere also transfers mechanical energy. The unequal heating of Earth's surface produces winds. Winds can be global, like the prevailing westerlies in the lower atmosphere that steer air masses and hurricanes. Winds are also local, like the

Figure 5.6-2. The water cycle.

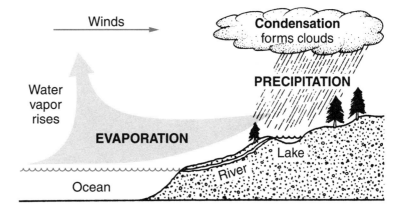

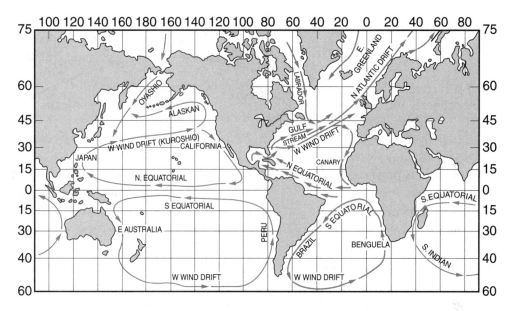

Figure 5.6-3. Ocean currents are produced by global winds that blow steadily in zones across the ocean.

land and sea breezes that you experience at the beach.

Winds blowing across water produce waves. Mechanical energy is transferred from the wind to the water's surface. The size of the wave depends on the speed of the wind, the distance the wind blows, and how long the wind lasts. The waves can vary in size from a ripple less than a centimeter in height to gigantic waves greater than 30 meters high. If you observe a calm lake, you may watch small waves develop when a breeze starts to blow across the surface.

Steady, global winds blowing across the ocean's surface produce horizontally flowing rivers of ocean water. These are called **ocean currents**. Between 30° and 60°N latitude, the prevailing westerlies blow from west to east. These winds produce an ocean current, called the Gulf Stream, that flows across the Atlantic Ocean from North Carolina to the British Isles. (See Figure 5.6-3.)

Activities

1. Toss three rocks into a pond or pool of water. The rocks should be about 1 cm, 3 cm, and 6 cm in size. Toss each with about the same force. Observe the size of the ripples produced. Basically, you are observing the transfer of mechanical energy from the rock to the water.

a) Compare the size of the rocks to the size of the waves produced.

b) Which rock produced the largest waves?

c) Which rock transferred the greatest amount of mechanical energy?

2. You can demonstrate the relationship between wind speed and wave height by blowing across a pan containing water. First, blow gently, and see what type of wave is produced. Compare this to the height of the waves when you blow more strongly, or when you use a fan to produce a breeze. What conclusion can you make regarding wind and waves? **(Caution: Use a low setting on the fan.)**

SKILL EXERCISE—*Designing a Model*

A model is a small version of some larger object. You may be familiar with model planes or cars. A model can also be designed to demonstrate large-scale events or processes so they can be studied or observed more closely. The figure below shows a model of how evaporation, condensation, and precipitation occur in our atmosphere.

Compare this figure with Figure 5.6-2 on page 195 and answer the following questions:

1. Where are evaporation, condensation, and precipitation taking place in the model

2. The sun provides the energy for evaporation in the atmosphere. What provides the energy in the model?

3. Will all the water in the kettle eventually be in the pan? Explain.

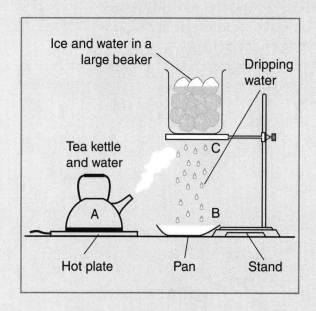

Questions

1. Oceans, glaciers, lakes, and rivers are part of Earth's
   (1) atmosphere   (3) hydrosphere
   (2) hemisphere   (4) lithosphere

2. An example of mechanical energy transfer between the atmosphere and hydrosphere is
   (1) waves   (3) clouds
   (2) condensation   (4) evaporation

3. The evaporation and condensation processes
   (1) transfer heat from the atmosphere to the hydrosphere
   (2) transfer heat from the hydrosphere to the atmosphere
   (3) absorb heat from the atmosphere
   (4) reflect radiant energy into space

4. The water cycle constantly moves water back and forth between the
   (1) oceans and land
   (2) lithosphere and atmosphere
   (3) atmosphere and lithosphere
   (4) hydrosphere and atmosphere

5. On a clear summer day you may observe waves at a local pond. The waves are most likely produced by
   (1) fish swimming in the water
   (2) wind blowing across the surface of the water
   (3) land vibrations
   (4) people throwing pebbles in the water

Thinking and Analyzing

1. Explain what is meant by the statement, "Without the water cycle we would eventually run out of freshwater on Earth."

2. Winds produce waves by causing friction at the water's surface. The figure below shows the relationship of waves and the factors that produce waves of different heights.
   a. What three factors influence the height of waves?
   b. What is the relationship between each of the factors and the size of waves?

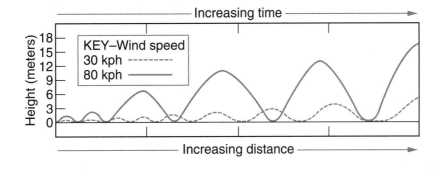

5.7 How Do Large Air Masses Affect Our Weather?

Objectives

Describe how an air mass forms.

Explain the type of weather associated with an air mass.

Terms

air mass: a large body of air that has generally the same temperature and humidity conditions throughout

continental (kahn-the-NEN-till) **air mass:** a dry air mass that forms over land

maritime (MEHR-eh-time) **air mass:** a moist air mass that forms over a large body of water

tropical (TRAHP-ih-kuhl) **air mass:** a warm or hot air mass that forms south of the United States.

polar air mass: a cool or cold air mass that forms north of the United States

Air Masses

A large body of air that has generally the same temperature and moisture conditions throughout is called an **air mass.** Specific weather conditions are associated with each type of air mass.

Air masses develop when air lingers over a large area of Earth's surface and takes on the characteristics of the temperature and moisture from the land or water under it.

The Names and Characteristics of Air Masses

Air masses are named for the temperature and moisture conditions of the region over which they form. **Continental air masses** form over land and are dry. **Maritime air masses** form over water and are moist. **Tropical air masses** form south of the United States and are warm. **Polar air masses** form north of the United States and are cool.

An air mass that forms over Canada is dry because it forms over land. It is also cool because it forms north of the United States. This air mass is called continental polar, or cP. The weather map abbreviation is lowercase "c" for continental, and uppercase "P" for polar.

Stop and Think: What two letters would you use to identify an air mass that formed in the Gulf of Mexico?

Answer: The weather map letters would be mT for maritime tropical.

Figure 5.7-1. There are six major types of air masses that affect the weather in the continental United States.

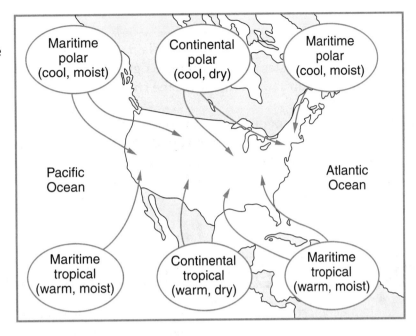

How Air Masses Affect Our Weather

The major air masses that affect the continental United States, shown in Figure 5.7-1, enter the country from the north, west, and south. The prevailing westerlies generally move major air masses across the United States from west to east.

As an air mass moves across the United States, it changes local weather conditions. The weather may become warmer or cooler, or wetter or drier, depending on the type of air mass passing over the region. For example, a maritime tropical air mass that forms in the Gulf of Mexico brings warm moist air to New York. (See Figure 5.7-2.)

Stop and Think: What type of air does a maritime polar (mP) air mass bring to New York?

Answer: A maritime polar air mass brings moist cool air to New York.

The surface area under a moving air mass also causes an air mass to change. A cP air mass moving across the United States will become warmer. Usually, the longer it takes to move, the warmer it will become.

Figure 5.7-2. A maritime tropical air mass brings warm, moist air to New York.

Activities

1. Obtain a map of the United States. Explain to someone (a parent, brother, sister, your schoolmate right next to you, or a friend) the six air masses that strike the United States. (See Figure 5.7-1.)

2. Describe the present outdoor weather conditions in terms of temperature and moisture. From your description, guess what type of air mass is affecting your area. Check the weather map in the newspaper or on TV to see if you are correct.

SKILL EXERCISE—*Creating a Chart*

*I*nformation can be put into an orderly form by creating a chart or a table. For instance, the table below represents information about air masses. Copy the chart on a separate piece of paper and fill in the missing information in each column.

Name of Air Mass	Map Abbreviation	Temperature (Cool or Warm)	Moisture (Moist or Dry)	Direction It Enters the US
Continental Polar				
Maritime Polar				
Continental Tropical				
Maritime Tropical				

Questions

1. An air mass is a large body of air that has about the same
   (1) temperature and moisture conditions throughout
   (2) moisture and pressure conditions throughout
   (3) temperature and pressure conditions throughout
   (4) temperature and precipitation conditions throughout

2. A dry air mass is most likely to form over
   (1) the Gulf of Mexico
   (2) the Pacific Ocean
   (3) the Atlantic Ocean
   (4) Canada

3. Air masses that strike southern California commonly form over the South Pacific Ocean. These air masses are
   (1) warm and dry (3) warm and moist
   (2) cool and dry (4) cool and moist

4. Which factor has the greatest influence on the movement of an air mass?
   (1) ocean currents
   (2) prevailing westerlies
   (3) sea breezes
   (4) mountain barriers

5. If, on January 15th, the people of New York City are having a sunny, but very cold day, we are probably in a
   (1) cP air mass   (3) cT air mass
   (2) mP air mass   (4) mT air mass

6. The map below shows an air mass that formed over the Gulf of Mexico at location *A*.
   Once air mass *A* reaches location *B*, the weather conditions at location *B* will most likely become
   (1) warmer and drier
   (2) warmer and more humid
   (3) colder and more humid
   (4) colder and drier

Thinking and Analyzing

1. Sometimes, in winter, an air mass flows into the United States from the Arctic region north of Canada. This air mass is labeled cA, which represents a continental arctic air mass. What are the temperature and moisture characteristics of a cA air mass?

2. As an air mass moves from west to east across the United States, its temperature and moisture characteristics change. Describe how a maritime polar air mass that enters the northwestern United States changes by the time it reaches New York City.

5.8

What Is a Weather Front?

Objectives

Explain how a weather front is formed.

Describe the weather produced along each type of front.

Terms

front: the boundary between different air masses

stationary (STAY-shah-nair-ee) **front:** the boundary produced when a cool air mass and a warm air mass meet and neither is capable of pushing the other

cold front: the boundary produced when cold air pushes under warm air

warm front: the boundary produced when warm air pushes up and over cooler air

occluded (uh-KLOOD-ed) **front:** the boundary produced when a cold air mass pushes cool air up and over warm air

Formation of a Front

When one air mass comes into contact with another air mass, a boundary, called a **front**, forms between them. The air masses that meet often differ in temperature, humidity, and density. These differences prevent the air masses from mixing. Cooler, drier air is denser and remains close to the ground. Warmer, moister air is less dense and is forced upward by the cooler, denser air. The rising air forms a low-pressure area along the front, often producing stormy weather conditions.

As discussed in the next sections, four different types of fronts can form, depending upon the motion of the air masses.

Types of Fronts

1. **Stationary Front:** When a cool air mass and a warm air mass come in contact, but neither air mass pushes the other, a stationary front forms. Along the front, there are usually clouds and some light precipitation. When one air mass starts pushing the other air mass, a different type of front is produced (see the following discussion). The map symbol for a stationary front is a line with arrowheads on one side and half circles on the other side. (See Figure 5.8-1.)

Figure 5.8-1. A stationary front is produced when a cold air mass and a warm air mass make contact and neither air mass pushes the other.

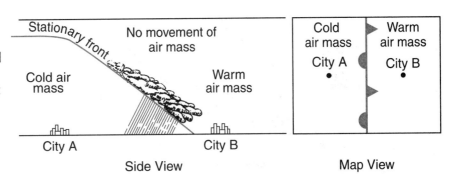

2. **Cold Front:** When a cold air mass pushes into and under a warm air mass, a cold front forms. Cold fronts usually bring heavy precipitation, cumulus cloud development, gusty winds, and cooler temperatures. After the cold front passes, the temperature and humidity usually drop. The map symbol for a cold front is a line with arrowheads on the side that shows the direction the front is moving (from cold air to warm air).(See Figure 5.8-2.)

3. **Warm Front:** When a warm air mass pushes into and over a cold air mass, a warm front forms. The warm air slides up and over the cooler air. Warm fronts usually bring stratus clouds and light precipitation that last a few days. After the warm front passes, the temperature and the humidity usually increase. The map symbol for a warm front is a line with half circles that shows the direction the front is moving (from warm to cold air). (See Figure 5.8-3.)

Figure 5.8-2. A cold front forms along the leading edge of a forward-moving cold air mass.

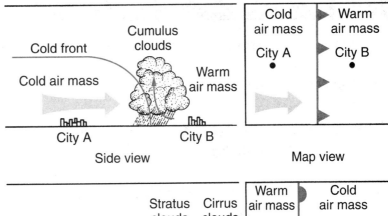

Figure 5.8-3. A warm front forms along the leading edge of a forward-moving warm air mass.

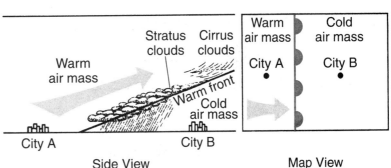

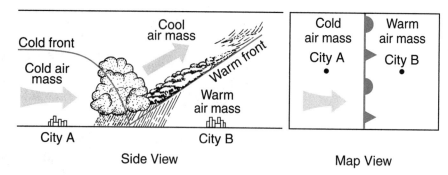

Figure 5.8-4 An occluded front forms when a cool air mass is squeezed upward between a fast-moving cold front and a slow-moving warm front.

4. **Occluded Front:** An occluded front forms when cool air is trapped and squeezed upward between a fast-moving cold front and a slow-moving warm front. Occluded fronts pass slowly. Stormy weather conditions are produced. The temperature changes from cool to warm to cold as the three air masses pass. The map symbol for an occluded front is a line with arrowheads and half circles on the same side that shows the direction the front is moving. (See Figure 5.8-4.)

SKILL EXERCISE—*Analyze a Weather Diagram*

*T*he figure below shows a cross section of a warm front moving from west to east.

1. Describe the present weather at location *C.*

2. Compare the temperatures at locations *A* and *D.*

3. If the front moves east at 400 km per day, what would the weather be like at location *D* the next day?

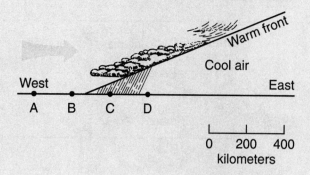

Activity

Collect weather maps for three consecutive days from a newspaper or the Internet. Look at the first day's weather map.

(a) What type of fronts can you find on the map?

(b) Select a city on each side of the front and describe the temperature of that city.

(c) What type of weather is associated with each front?

(d) During the three days, how did the fronts move?

Questions

1. The diagram below shows two symbols commonly found on a weather map. The symbols on this map represent
   (1) winds
   (2) fronts
   (3) latitude and longitude
   (4) climatic conditions

2. Fronts are produced at the boundary of air masses. What type of weather is usually associated with the passing of a cold front?
   (1) fair, clear, and warmer
   (2) cloudy, rainy, and colder
   (3) cloudy, rainy, and warmer
   (4) fair, cloudy, and colder

3. What type of air pressure develops along fronts?
   (1) low air pressure (3) high air pressure
   (2) rising air pressure (4) no air pressure

Thinking and Analyzing

1. The table at the right shows a list of the temperature and air pressure for five days in New York City.
   (a) On what day did a front pass?
   (b) What type of front was it?
   (c) Describe what type of weather there probably was when the front passed.

2. When a mT and a cP air mass meet, a front is produced. Why don't the two air masses mix?

Date	Temperature °F	Temperature °C	Air Pressure (in. of mercury)
Oct 14	62°	17°	30.30
Oct 15	64°	18°	30.25
Oct 16	42°	6°	29.86
Oct 17	40°	4°	30.05
Oct 18	50°	10°	30.20

What Is a Hurricane?

Objectives

Describe the characteristics of a hurricane.

Explain how a hurricane develops and the typical path it takes.

Terms

hurricane (HUR-ih-kayn): a large tropical storm with rotating wind speeds greater than 74 mph

storm surge: a wall of ocean water pushed toward land during a hurricane

tropical depression: a storm formed over warm water with rotating wind speeds less than 39 mph

tropical storm: a storm formed over warm water with rotating wind speeds between 39 and 74 mph

---

Characteristics of a Hurricane

Hurricanes are large rotating tropical storms that develop over warm ocean water north of the equator. They form in an area extending from the west coast of Africa to the Gulf of Mexico. A storm is classified a hurricane when its wind speeds reach 74 mph or greater. (See Figure 5.9-1.)

A fully developed hurricane can have a diameter of 400-500 miles. Winds rotate counterclockwise around a calm central region, called the eye. A well-formed eye can be 25–50 miles in diameter. Wind speed is greatest around the eye, and decreases as it moves outward from the eye. Columns of upward-developing clouds form a wall around the eye of the hurricane. These clouds sometimes reach heights greater than 6 miles. The air pressure in the eye of a hurricane can be as low as 28.00 inches of mercury. (See Figure 5.9-2.)

Figure 5.9-1. A satellite view of Hurricane Katrina (2005) striking the southeast coast of the United States.

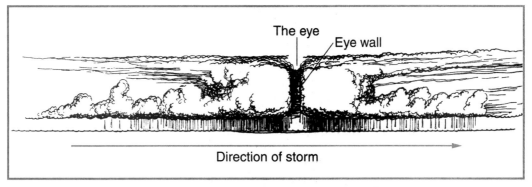

Figure 5.9-2. A cross section of a hurricane showing the cloud structure, the eye, and the eye wall.

Strong winds, heavy rain, and large powerful waves can cause considerable damage, especially when a hurricane strikes land along the coastline. When a wall of ocean water, called a **storm surge**, moves toward land ahead of a hurricane, it often causes great devastation to shoreline structures, especially at high tide.

Hurricane Katrina (2005) was the most damaging hurricane to ever hit the United States. The storm was greater than 500 miles wide, and had surface winds greater than 175 mph. Air pressure dropped to under 27 inches of mercury. A storm surge of about 28 ft hit the coast of Louisiana. More than 1500 people died, and total damage was estimated at more than $60 billion.

The Path of a Hurricane

To understand the typical path of a hurricane, you should understand the forces that steer them:

- A global wind belt in the lower atmosphere, called the northeast trade winds, blows from east to west between 0°–30° N latitude.
- Another global wind belt in the lower atmosphere, called the prevailing westerlies, blows from west to east between 30°–60° N latitude.
- Earth's rotation causes moving objects in the northern hemisphere to curve to the right.
- High air pressure systems tend to block the movement of hurricanes.

Hurricanes travel a variety of paths. However, most hurricanes that strike the United States follow a typical path.

Hurricanes usually develop between 5°–20° N latitude in the Atlantic Ocean, Caribbean Sea, or Gulf of Mexico. At these low latitudes, the northeast trade winds push hurricanes to the west. The rotating Earth causes hurricanes to curve to the right, or northward. Frequently, hurricanes strike the southeast coastline of the United States. (See Figure 5.9-3 on the next page.)

Stop and Think: Which states in the United States are most affected by hurricanes?

Answer: The coastal states from Texas to North Carolina are most affected by hurricanes.

As the storm moves north, the prevailing westerlies turn the hurricane eastward. High-

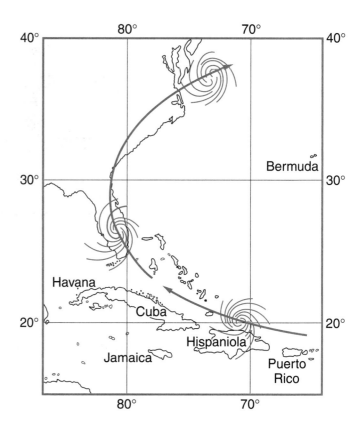

Figure 5.9-3. A typical path of a hurricane striking the United States.

pressure systems from Canada and the Gulf of Mexico may block its forward motion in the middle latitudes. The storm usually moves into the cooler waters of the North Atlantic Ocean and dies.

The Life Cycle of a Hurricane

Hurricanes develop from June to November when the ocean water is warmest. A hurricane begins as a weak weather disturbance over tropical waters. As rotational wind speed increases, the storm grows from a **tropical depression** (winds less than 39 mph) to a **tropical storm** (winds between 39–74 mph), and eventually becomes a hurricane (winds greater than 74 mph). Powerful hurricanes can have wind speeds greater than 150 mph. Hurricanes gain energy over warm ocean

waters from the evaporation and condensation occurring in massive cloud formations. Table 5.9-1 shows the five hurricane categories based upon wind speed, air pressure, storm surge, and damage.

When a hurricane strikes land, it loses its source of energy (the warm ocean water) and its wind speed quickly decreases. The storm may turn into a heavy rainstorm and cause severe inland flooding. As a hurricane curves to the northeast and its wind speed decreases, it is downgraded to a tropical storm or even a tropical depression. Eventually, the storm's circular structure breaks apart as it moves further inland or over cooler ocean water.

On August 23, 2005, Hurricane Katrina developed quickly in the

Table 5.9-1. Hurricane Categories

Category	Wind Speed (mph)	Air Pressure (inches of Mercury)	Storm Surge (ft)	Damage
1	74–95	28.94 or more	4–5	Tree branches and power lines down
2	96–110	28.50–28.93	6–8	Large signs and small trees blown down
3	111–130	27.91–28.49	9–12	Small buildings and large trees blown down
4	131–155	27.17–27.90	13–18	Most roofs, walls, and windows destroyed
5	155 or more	less than 27.17	18 or more	All buildings and large structures destroyed

Caribbean Sea. In two days, it became a hurricane and moved west. It crossed the southern tip of Florida and moved into the Gulf of Mexico. It made a second landfall in Louisiana on August 29. After striking New Orleans, it moved northward and lost strength, but it still caused considerable inland flooding. By September 1, Hurricane Katrina had become a minor storm in the Northeast.

Activity

Obtain a copy of the Atlantic Basin Hurricane Chart on the Internet at *www.nws.noaa.gov* (follow the links to hurricanes). Track the daily position of a current hurricane. Get the daily latitude and longitude location from a TV weather forecast, an Internet site, or a newspaper. If there is no current hurricane activity on the Internet at the NOAA site, obtain data for the path of a hurricane that affected the United States within the last year.

SKILL EXERCISE—*Using Latitude and Longitude Map Skills*

*T*he Atlantic Basin Hurricane Tracking Chart shows the paths of Hurricanes Floyd (1999)—dashed—and Katrina (2005)—solid. Each dot represents a 1-day change in position of the hurricane's path.

Atlantic Basin Hurricane Tracking Chart
National Hurricane Center, Miami, Florida

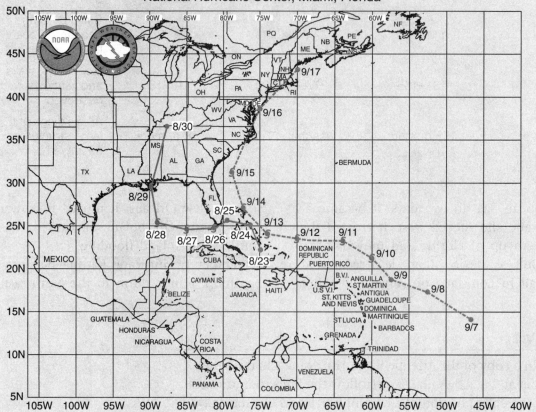

Study the map carefully and answer the following questions:

1. Did Hurricane Floyd take a typical hurricane path? Explain.

2. What was Floyd's latitude and longitude on September 7 (9/7)?

3. On August 29, Hurricane Katrina had a wind speed of 125 mph. On August 30, the wind speed decreased to 30 mph. What caused the decrease in wind speed?

4. What is the most likely location of Hurricane Katrina on August 31 (8/31)?

(1) 35° N, 75° W

(2) 40° N, 95° W

(3) 35° N, 65° W

(4) 40° N, 80° W

Interesting Facts About Hurricanes

Hurricanes occur in the North Atlantic Ocean. Similar storms form around the world, but are given different names. When storms form In the Pacific Ocean, they are called typhoons. When storms form in the Indian Ocean they are called cyclones. Storms that form around Australia are sometimes called willy-willies.

The worst natural disaster in the United States, based on total deaths, was caused by a hurricane that struck Galveston, Texas, in 1900. It killed 6000 people. Worldwide disasters connected to hurricanes, typhoons, cyclones, and willy-willies have been very costly to human life. The table below lists the tragic results of four hurricane-type storms.

Location	Year	Deaths
Calcutta, India	1737	300,000
Indochina (SE Asia)	1881	300,000
Bangladesh	1970	500,000
Bangladesh	1991	139,000

Questions

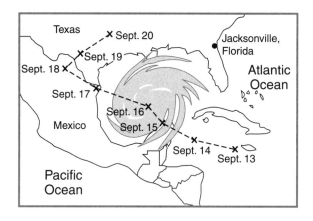

1. The map at the right shows the relative size and track of a certain weather event. Which weather event is shown on the map?
   (1) tornado   (3) hurricane
   (2) blizzard   (4) thunderstorm

2. Hurricanes are about the size of
   (1) New York City
   (2) Texas
   (3) New York State
   (4) the Atlantic Ocean

3. In the center of a hurricane there is a calm area called the
   (1) eye   (3) wall
   (2) nose   (4) depression

4. The path of a hurricane is mostly controlled by
   (1) the coastline
   (2) wind belts in the lower atmosphere
   (3) ocean currents
   (4) land

5. The source of energy that produces a hurricane comes from
   (1) warm ocean water
   (2) directly from the sun
   (3) clouds
   (4) ocean currents

6. Shown below, the typical path of a hurricane is illustrated in Figure
   (1) (1)   (3) (3)
   (2) (2)   (4) (4)

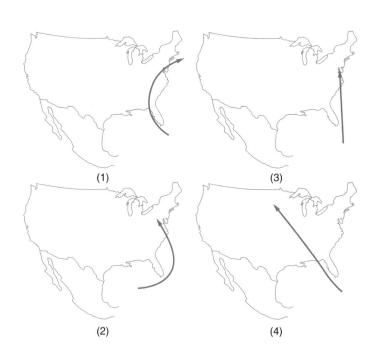

Thinking and Analyzing

1. The figure at the right shows the changing wind speed within a hurricane. Carefully review the graph and answer the following questions:

   a. Where is the wind speed the greatest in a hurricane?

   b. Where is the wind speed the least in a hurricane?

   c. Describe how the wind speed changes as you move outward from the eye of a hurricane.

2. Almost all hurricanes develop between June and November. These months are called the hurricane season. Explain why hurricanes don't form in January.

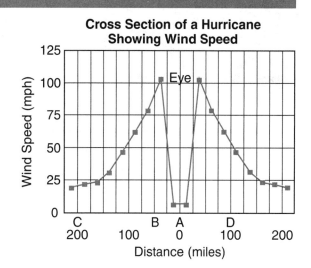

Cross Section of a Hurricane Showing Wind Speed

5.10

What Is a Tornado?

Objectives

Describe the characteristics of a tornado.

Explain how a tornado forms.

Terms

tornado (tawr-NAY-doh): a violently spinning funnel-shaped cloud

thunderstorm: a storm with lightning and thunder, very gusty winds, heavy rain and sometimes hail

tornado alley: an area, from Texas to Indiana, where a high number of tornadoes occur

---

Characteristics of Tornadoes

Tornadoes are violently whirling winds, shaped like a funnel. The funnel extends down from a cloud and often touches the ground. (See Figure 5.10-1.) Spiraling, high-speed winds, usually greater than

Figure 5.10-1. A tornado funnel extending from cloud to ground.

200 mph (320 kph), are strong enough to lift cars and even trains. Tornadoes are most commonly produced within violent thunderstorms, but many also form in hurricanes.

Tornadoes are also called twisters. They usually appear suddenly and carve a narrow path of destruction, They disappear as suddenly as they arrived. An average tornado travels about 45 kph (about 30 mph). It cuts a path about 150 to 600 meters wide and 25 km (15 miles) long.

The Development of a Tornado

Most tornadoes form within **thunderstorms** between the months of March and July. In the spring, cool dry, continental polar air moves south from Canada, and warm moist, maritime tropical air moves north from the Gulf of

Mexico. When these air masses collide, strong updrafts produce thunderstorms with large cumulonimbus clouds. Lightning, thunder, strong gusty winds, and sometimes hail are characteristics of a thunderstorm. The typical thunderstorm lasts a few hours.

Sometimes cool air is trapped above warm air. The dense, cool air wants to sink, and the less dense, warm air wants to rise. The exchange of these two types of air produces a whirling funnel cloud. The funnel begins in the cloud and moves down to the ground. (See Figure 5.10-2.) It may touch the ground for a few hundred meters, or a few hundred kilometers, before it lifts back into the cloud. The narrow path of destruction is often devastating and unpredictable. On a single street, a tornado might completely destroy one house, cause minor damage to the one next door, and leave the next house completely untouched.

Tornado Alley

More than 700 tornadoes strike the United States each year. Texas, Oklahoma, Kansas, Missouri, Nebraska, Iowa, Illinois, and Indiana have the greatest tornado risk. These states stretch across the Midwestern United States. It is within this area that air masses from the north and south usually collide each spring. This area is commonly referred to as **tornado alley** because of the high number of tornadoes that form there. Some regions of Mississippi, Alabama, and Florida also have a high number of tornadoes. (See Figure 5.10-3 on the next page.)

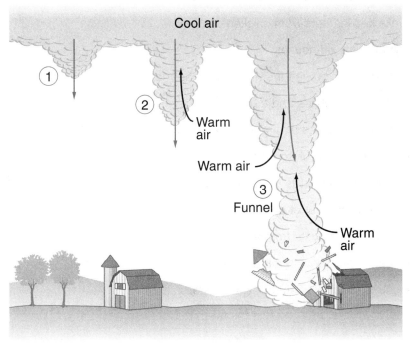

Cool air

1

2

Warm air

Warm air

3

Funnel

Warm air

Figure 5.10-2. Stages in the development of a tornado: (1) cool, dense air sinks, (2) warm air whirls in from the sides, and (3) the funnel extends down to the ground.

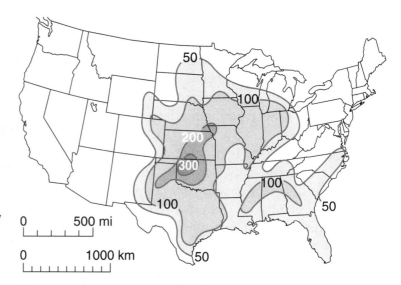

Figure 5.10-3. Most tornadoes form in the midwest where cold, dry air from Canada collides with warm, moist air from the Gulf of Mexico.

Interesting Facts About Tornadoes

Wind speeds in a tornado may reach 250–300 mph (400–480 kph). In addition to great wind speed, tornadoes have extremely low pressure that causes powerful updrafts. Here is a list of some odd but destructive events that tornadoes have produced.

Tornadoes have:

- driven stalks of wheat into trees and utility poles
- plucked feathers from chickens
- lifted roofs off houses
- lifted barns, but left the animals unharmed
- lifted automobiles and trains
- lifted and carried farm animals and placed them down unharmed
- sheared the wool from sheep
- driven wood boards through steel girders
- sucked soda out of open bottles
- pulled fence posts from the ground
- lifted a dresser with mirrors and placed it down without damage
- caused it to rain fish, frogs, and snails

SKILL EXERCISE—*Interpreting Map Data*

*T*he map of the United States below shows the annual average number of tornadoes in each state for the years 1950–1995. Although the annual average number of tornadoes in each state gives an indication of where most tornadoes occur, it does not indicate which state has the greatest risk of a tornado striking your house. To determine the risk of a tornado striking you, the size of the state in which you live must be taken into account. For example, New York State averaged 6 tornadoes per year and New Jersey averaged 3 tornadoes per year, but the personal risk is greater in New Jersey because it is a smaller state.

Carefully study the map and answer the following questions:

1. Which state had the highest annual average number of tornadoes for the 45-year time period?

2. What was the annual average number of tornadoes per year for Oklahoma?

3. There was a greater risk of a tornado striking your house in Oklahoma than in Texas. Explain why.

4. Why do most tornadoes occur in the states between Texas and Illinois?

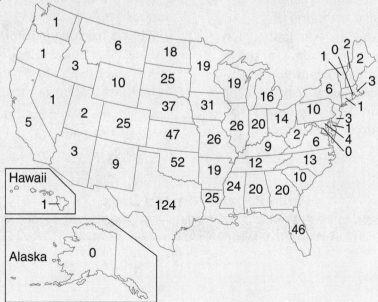

Annual Average Number of Tornadoes, 1950–1995

Questions

1. Tornadoes are associated with
   (1) only hurricanes
   (2) only thunderstorms
   (3) hurricanes and thunderstorms
   (4) large air masses

2. The violent windstorm visible as a funnel-shaped cloud is called a
   (1) hurricane   (3) tornado
   (2) thunderstorm   (4) blizzard

3. Tornadoes are most common in what part of the United States?
   (1) western states
   (2) midwestern states
   (3) eastern states
   (4) southeastern states

4. Compare the size of a tornado to a hurricane and a thunderstorm. Which statement is correct?
   (1) A tornado is smaller than both a thunderstorm and a hurricane.
   (2) A tornado is larger than both a thunderstorm and a hurricane.
   (3) A tornado is larger than a thunderstorm, but smaller than a hurricane.
   (4) A tornado is the same size as a thunderstorm, but smaller than a hurricane.

5. During what months do most tornadoes form?
   (1) March to July
   (2) June to November
   (3) January to March
   (4) June to September

Thinking and Analyzing

1. The table below shows the monthly percentage for the distribution of tornado days during the year. January has 2.1 percent of the tornado days for the year. This low number means that few tornadoes form in January. May has 22.9 percent of tornado days. This means that more tornadoes occur in May than in any other month.

   a. What months make up the tornado season?
   b. What is the total percentage of tornado days in the tornado season?
   c. Why does the highest percentage of tornado days occur in these months?

2. Compare the size, duration, and wind speed of a tornado with a hurricane and thunderstorm.

Percentage of Tornado Days

Jan	Feb	Mar	Apr	May	Jun	Jul	Aug	Sep	Oct	Nov	Dec
2.1	3.4	9.8	15.7	22.9	18.2	9.8	5.5	4.7	2.5	3.4	2.0

Review Questions

Term Identification

Each question below shows two terms from Chapter 5. One of the terms is defined.
(1) Choose the term that matches the definition.
(2) Describe how the two terms are different. Following each term is the section (in parenthesis) where the description or definition of that term is found.

1. *Evaporation (5.4) — Condensation (5.5)*
   The process that occurs when liquid water changes into water vapor; a cooling process that takes heat from its surroundings

2. *Cold front (5.8) — Warm front (5.8)*
   The front that forms when warm air pushes up and over cool air

3. *Hurricane (5.9) — Tornado (5.10)*
   A large storm with winds greater than 74 mph and a calm eye in the center

4. *Maximum humidity (5.4) — Relative humidity (5.4)*
   A ratio (%) between the actual amount of water vapor in the air and the maximum humidity the air can hold

5. *Air mass (5.7) — Front (5.8)*
   A large body of air that takes on the temperature and moisture characteristics of Earth's surface beneath it

6. *Hydrosphere (5.6) — Atmosphere (5.6)*
   Consists of gases that surround Earth

7. *Hail (5.5) — Sleet (5.5)*
   The type of precipitation that is produced by freezing rain; occurs often in winter months

8. *Maritime (5.7) — Continental (5.7)*
   Represents a moist air mass that forms over the ocean

Multiple Choice (Part 1)

Choose the response that best completes the sentence or answers the question.

1. The atmosphere receives heat by (a) absorbing solar radiation, (b) absorbing radiant heat from Earth, and (c)
   (1) direct contact with Earth
   (2) winds blowing over the ocean
   (3) melting of glaciers
   (4) earthquake activity

2. Air pressure is
   (1) produced by wind
   (2) the weight of the overlying air
   (3) caused by an interaction of the hydrosphere and atmosphere
   (4) produced by radiant energy from the sun

3. The column of mercury in a mercury barometer
   (1) is produced by the expansion of the mercury
   (2) is determined by the amount of mercury in the dish
   (3) remains constant when left in one place
   (4) is balanced by the force of air pushing down on the mercury in the dish

4. A rising barometer indicates fair weather because a
   (1) high-pressure system is approaching
   (2) low-pressure system is approaching
   (3) a cold front is approaching
   (4) a warm front is approaching

5. Relative humidity is a ratio between
   (1) the wet-bulb and dry-bulb temperature readings
   (2) the actual temperature and maximum temperature
   (3) actual humidity and maximum humidity
   (4) the humidity in two colliding air masses

6. Relative humidity is measured with a wet-bulb and dry-bulb thermometer. What information do you need to find the relative humidity?
   (1) the wet-bulb temperature
   (2) the dry-bulb temperature
   (3) the dry-bulb temperature and the difference between the wet- and dry- bulb temperatures
   (4) the sum of the wet- and dry-bulb temperatures and the dry-bulb temperature

7. Precipitation forms within a cloud when
   (1) strong downdrafts push water droplets to the ground
   (2) ice crystals melt on their way to Earth
   (3) water vapor changes into water droplets
   (4) water droplets or snow crystals become too heavy and fall to Earth

8. Which type of precipitation is frozen rain?
   (1) sleet
   (2) hail
   (3) snow crystals
   (4) snowflakes

9. Which process is an example of mechanical energy transfer from the atmosphere to the hydrosphere?
   (1) evaporation from the ocean and condensation in the atmosphere
   (2) heat exchange by direct contact of the ocean with the atmosphere
   (3) winds blowing across the ocean and producing waves
   (4) underwater earthquakes producing giant waves

10. Heat energy from the ocean is transferred to the atmosphere by the processes of
    (1) precipitation and condensation
    (2) evaporation and condensation
    (3) precipitation and evaporation
    (4) heating and cooling

11. A stationary air mass in the Atlantic Ocean off the east coast of Florida would have what characteristics?
    (1) warm and dry
    (2) cool and dry
    (3) warm and moist
    (4) cool and moist

12. What type of weather is associated with a continental polar (cP) air mass?
    (1) cold and stormy
    (2) cold and fair
    (3) warm and stormy
    (4) warm and fair

13. A cold front is formed
    (1) along the leading edge of warm air pushing over cold air
    (2) along the boundary between two stationary air masses
    (3) when cold air slides up and over a warm air mass
    (4) along the leading edge of cold air pushing under warm air

14. Weather along a front is most often
    (1) fair and sunny
    (2) fair and cloudy
    (3) cloudy and stormy
    (4) cloudy and cold

15. Hurricanes are
    (1) large storms, containing an eye, with winds over 74 mph
    (2) large storms, containing a funnel cloud, with winds over 200 mph
    (3) large storms, containing thunder, lightning, and hail
    (4) small storms, containing thunder, lightning, and a funnel cloud

16. Hurricanes that form north of the equator in the Atlantic Ocean commonly strike
    (1) the midwestern states.
    (2) the New York City area
    (3) California
    (4) the southeastern states.

17. The most unique characteristic of a tornado is
    (1) thunder and lightning
    (2) a funnel cloud
    (3) a calm eye in the center
    (4) strong winds and blowing snow

18. Most tornadoes that form in the United States form where continental polar air masses and maritime tropical air masses clash. This section of the United States is
    (1) the western states
    (2) the eastern states
    (3) the midwest
    (4) the Rocky Mountain States

Thinking and Analyzing (Part 2)

1. A continental polar air mass forms over Canada and moves south to Kentucky, as shown in the figure below.

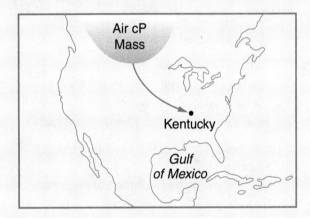

(a) What type of front will affect the weather of Kentucky before the air mass arrives?
(b) Describe the type of weather the front will most likely bring Kentucky.
(c) Describe the type of weather the air mass will most likely bring Kentucky.

2. Use the August weather map of New York State below and the following information to answer parts (a) and (b).

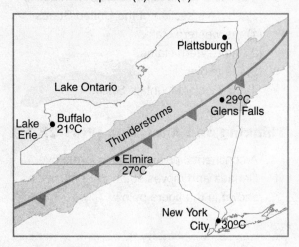

Current weather conditions are:

Buffalo – clear skies, cool temperature
Glens Falls – windy, thunderstorms
Elmira – windy, thunderstorms
New York City – clear skies, hot and humid

a. What is the most likely weather in Plattsburgh?
b. Predict what the weather will most likely be in New York City within the next 6 hours.

3. The map below represents a satellite image of Hurricane Gilbert in the Gulf of Mexico. Each X represents the position of the center of the storm on the date indicated. Compare the strength of Hurricane Gilbert on September 16 and September 19. What was the cause of the difference in strength?

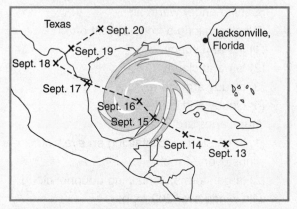

4. The map of the United States below shows the distribution of tornadoes over a 12-year period.
(a) What is the average number of tornadoes during the 12 years that affected Florida, New York, and Illinois?
(b) Describe where in the United States most tornadoes occur.
(c) Why do most tornadoes occur in this area of the United States?

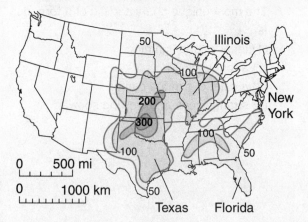

Chapter Puzzle (*Hint:* The words in this puzzle are terms used in the chapter.)

Across

5. warm air from south of US

10. _____ breeze: cool breeze that blows onshore in daytime

12. tropical _____; tropical storm with rotating winds less than 39 mph

13. a violently spinning funnel-shaped cloud

14. dry air mass that forms over land

15. moist air mass that forms over a large body of water

19. air _____; force of air pushing down on an object

20. barometer that does not contain liquid

21. form of energy capable of passing through space

24. puffy clouds

26. _____front; cold air pushes under warm air

27. mass per unit volume of a substance

31. state of the atmosphere

32. _____humidity; ratio of water vapor in the air and the maximum amount air can hold

34. large tropical storm with rotating winds greater than 74 mph

35. ice precipitation formed in layers

36. _____ humidity; greatest amount of water vapor the air can hold

Down

1. air _____; a large body of air with uniform temperature and humidity

2. storm associated with lighting

3. boundary between different air masses

4. water _____; gas state of water

6 liquid or solid water that falls to earth

7 flowing movement of air or water

8 wind _____ index; affect wind has on how we perceive cold

9 instrument to measure air pressure

11 _____ front; warm air squeezed between two cooler air masses

16 cool air from north of US

17 surface and subsurface water on Earth

18 change a gas to liquid

19 instrument used to measure relative humidity

22 tornado _____; area in the US where most tornadoes form

23 horizontal movement of air

25 frozen rain

28 flat, layered clouds

29 unit of air pressure used on weather maps

30 storm _____; wave of ocean water pushed in front of hurricane

33 type of front when warm air pushes up and over cooler air

Unit 3

Diversity of Life

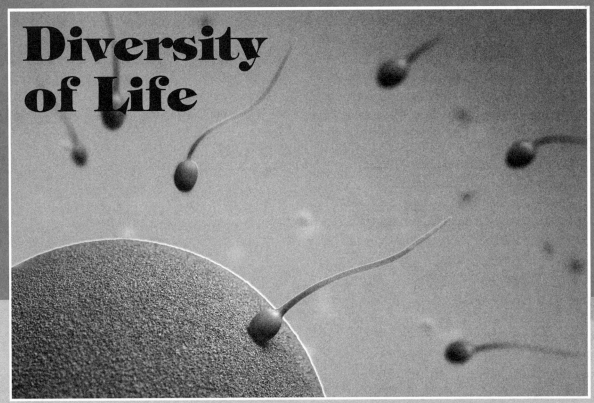

Part I—Essential Question

This unit focuses on the following essential question:

How does the transfer of matter and energy through biological communities support the diversity of living things?

In Unit 1, we studied physics. In Unit 2, we studied chemistry and Earth science and learned how matter and energy interact to produce weather patterns. In Unit 3, we are returning to the concepts of matter and energy, but this time we are studying matter and energy in living things. In this unit, we enter a fourth branch of science—biology, the study of life.

You already know that the sun is Earth's primary source of energy. It is also *our* primary source of energy. In Unit 3, you will learn how the energy from the sun is stored by green plants, and eventually transferred to animals like us. When we eat another animal, we obtain some of that animal's energy. When you eat beef you obtain energy from the cow. The cow got that energy from green plants. The plants got that energy from the sun.

All living things must obtain energy to survive. They do so in many different ways. The *diversity of living things* refers to the many different kinds of living things, and the many different adaptations that keep them alive.

Part II—Chapter Overview

Chapter 6 is called "Kingdoms of Life." The chapter focuses first on the similarities of living things. One thing that all organisms have in common is that they are made up of one or more cells, the basic units of life. Understanding what makes something alive, and what organisms must do to stay alive, will help you to appreciate the enormous variety of living things.

Scientists have classified organisms into five large categories, or kingdoms. A *kingdom* is a classification used by scientists to group together organisms that have certain similarities. In Chapter 6, we will investigate some of the structures that are typical of each kingdom, and how different organisms perform the same activities in different ways. For example, different organisms use different structures for *locomotion*, which is an organism's movement from place to place.

Chapter 7 is called "Food Chains and Food Webs." Food chains and food webs are diagrams that show the flow of energy from organism to organism within an environment. Each organism in the environment has a specific role to play. A giraffe obtains energy by eating the leaves of a tall tree, but if the giraffe is unlucky, it may end up supplying energy to a hungry lion.

In this chapter, we will see that living things also obtain oxygen and water from their environment. Living things have been consuming water and using oxygen for many thousands of years, yet the amount of water and oxygen on Earth has stayed about the same. How is this possible? In Lesson 7.2, you will learn how materials like oxygen and water are constantly recycled. Other materials, such as carbon and carbon dioxide, are recycled as well.

Chapter 6

Kingdoms of Life

Contents

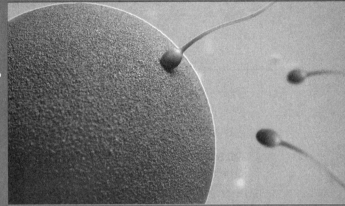

The picture above shows an egg cell, the largest cell in the human body, and sperm cells, the smallest cells in the human body.

What Is This Chapter About?

Biology is the study of life. What does a scientist mean by life? Living things range in size from whales to organisms too small to be seen without a microscope. Plants, animals, and bacteria are examples of living things. Living things, or organisms, share certain characteristics that set them apart from nonliving things. This chapter discusses some similarities and differences among all living things.

In this chapter you will learn:

1. Living things carry out *life functions* such as nutrition, respiration, and excretion.

2. The cell is the basic unit of living things.

3. Living things are organized into five different kingdoms.

4. Organisms adapt in different ways in order to survive.

Career Planning: Cell Biologist

Biologists are scientists who study living things. They have careers in research, animal care, and protecting wildlife, to name a few.

Elaine Fuchs (FYOOKS) is a cell biologist working at The Rockefeller University in New York City. She is studying skin and hair cells. She has discovered that there are stem cells in hair follicles. Stem cells are important because they have the potential to develop into other types of cells. Dr. Fuchs has discovered clues into what goes wrong when cells in the skin behave abnormally, and how disorders such as skin cancers develop. The work of cell biologists like Dr. Fuchs may provide a new source of stem cells, and future treatments or cures for illnesses.

Internet Sites:

http://www.biology4kids.com/map.html This Web site has many links to topics covered in this chapter, such as cell structures, cell functions, and kingdoms.

http://www.beyondbooks.com/lif71/4.asp Learn all about cells, from the origin of the cell theory to the structures within a cell.

http://www.tvdsb.on.ca/westmin/science/sbi3a1/Cells/cells.htm Click on the organelles and watch them in action.

6.1

What Makes Something Alive?

Objectives

Identify how living things are alike.

Describe life functions.

Terms

life functions: processes that occur in all living things

organism: a living thing

cell: the building block of living things

Living Things

Perhaps you have a green plant sitting on your windowsill. How do you know that the plant is alive? By observing the plant over time, you might notice some changes in the plant. The plant might be growing or bending toward the sunlight. These changes are examples of activities performed by living things. Growth and responding to changes in the surroundings (the sunlight) are two examples of processes called **life functions**. Living things are called **organisms.** Organisms must perform life functions to be considered alive. Table 6.1-1 lists and defines many of the life functions carried out by organisms. Don't be fooled! Some nonliving things do some of these activities as well. For example, a car releases the energy stored in fuel to make it move, yet a car is not alive.

Different types of organisms carry out these life processes in different ways. By

Table 6.1-1. The Life Functions of Organisms

Function	Purpose
Respiration	Release the energy stored in food
Transport	Move materials throughout the organism
Nutrition	Take in substances and/or break down food into a form cells can use
Synthesis	Manufacture new molecules
Excretion	Eliminate waste materials produced by the organism
Regulation	Respond to changes in the organism's surroundings
Locomotion	Move the organism from place to place
Reproduction	Make more organisms of the same kind (offspring)
Growth	Increase body size, replace damaged cells, or both
Development	Change from a young organism into an adult

Table 6.1-2 Life Functions Observed in a Plant.

Life Function	Observation
Nutrition	The plant has roots that take in water from the soil
Transport	The water moves through the stem to the rest of the plant
Growth	The stem, leaves, and roots get bigger
Regulation	If you place the plant in a sunny window, the plant bends toward the sunlight
Synthesis	Using the sunlight, a green plant can make its own food from water and carbon dioxide. (You will learn more about this process in Lesson 7.8.)

Figure 6.1-1. Humans eat to get food. Plants make their own from sunlight, water, and carbon dioxide.

observing life functions, we can easily determine that something is alive. Which of these can we observe in our green plant? (See Table 6.1-2.)

All of these life functions work together to keep the plant alive.

You can probably identify how most of these same life functions occur in your daily life. Although green plants can make their own food internally, you cannot! You must eat food that must then be digested (nutrition). Your blood moves the usable food through your body (transport). The oxygen you breathe in is needed to release the energy stored in the food (respiration). You blink, duck, and sneeze in response to changes in your environment (regulation). You are moving (locomotion), growing (growth), and maturing (development).

While you and your plant perform life functions in very different ways, you are meeting the same needs. All living things require energy, living space, and materials such as food, water, and oxygen.

Figure 6.1-1 shows a plant on a windowsill and a child eating lunch. The plant is turned toward the light. A plant uses sunlight to make its own food. Humans eat to get food.

Since the needs of all organisms are similar, it is no surprise that organisms share certain structures that permit them to meet these needs. These structures, called **cells**, are the basic building blocks of all living things. Cells are the topic of the next lesson.

Interesting Facts About Life

Have you ever had the flu? Perhaps you have had a flu shot to prevent the flu. This disease, like many others, is caused by a virus. A virus is a tiny collection of chemicals many times smaller than most cells. Viruses are not made up of cells and do not carry out life functions. Are viruses alive? Viruses invade the cells of living things and affect their activities, sometimes harming the host cell. Once inside the cell, viruses use materials within the cell to produce new viruses. In this way, viruses are performing one life function – reproduction. The new viruses then invade more cells. If enough cells are infected, the life functions of the infected organism may be damaged.

Does one life function make something alive? Scientists are still undecided as to whether to classify viruses as living things. What do you think?

Questions

1. Several tomato plants are grown indoors next to a sunny window. The plants receive water and fertilizer and remain on the windowsill. What will most likely happen?
   (1) Most of the leaves of the plants on the window side will wilt and die.
   (2) The roots of the plants will grow upward from the soil.
   (3) Water droplets will collect on the leaves of the plants facing away from the window.
   (4) The stems of the plants will bend toward the window.

2. Which life function releases energy stored in food?
   (1) nutrition   (3) excretion
   (2) respiration   (4) transport

3. Which statement is associated with the process of excretion?
   (1) The organism eliminates wastes to its surroundings.
   (2) The organism removes oxygen from its surroundings.
   (3) The organism takes in food from its surroundings.
   (4) The organism takes in carbon dioxide from its surroundings.

4. Which of the following life processes is **not** described correctly?
   (1) transport: moving materials throughout the body
   (2) regulation: responding to changes
   (3) reproduction: producing new individuals
   (4) nutrition: increase in body size

Thinking and Analyzing

1. Identify the life process described in each of the statements below.
   (1) Plants move water from the roots to the leaves.
   (2) Marathon runners digest the pasta that was eaten the night before to provide them with a source of energy.
   (3) You get "goose bumps" when it is cold.

2. List three ways in which you are similar to a plant. List three ways in which you are different.

3. You are trying to convince a friend that your car is similar to a living thing. Identify two processes that occur in a car, and explain why these appear to be life processes. Identify two life functions that your car cannot do.

What Are Cells?

Objectives

State the cell theory.

Describe how cells are the unit of structure of living things.

Describe how cells are the unit of function of living things.

Identify some cell structures and state their functions.

Terms

nucleus (NEW-clee-uhs): the structure within the cell that controls cell activities

cell membrane: the outer covering of the cell that controls the flow of materials into and out of the cell

cytoplasm (SYE-toe-plaz-uhm): the gel-like substance that fills the cell

organelle (or-guh-NEHL): a small structure within the cytoplasm that carries out a specific life function

unicellular (YOU-nih-SELL-you-luhr): consisting of a single cell

multicellular (MUHL-tee-SELL-you-luhr): containing more than one cell

photosynthesis (foh-toh-SIN-thuh-sis): the process through which plants use the energy in sunlight to produce food

---

Cell Theory

All living things are made of basic units called *cells*. Cells were first observed in 1665 when Robert Hooke used a new invention, called a microscope, to look at a piece of cork. Improvements in the microscope and the work of many scientists led to the development of the cell theory in 1885. The cell theory states that:

1. All living things are made up of cells.
2. Cells are the basic units of living things.

3. Cells come from previously existing cells.

If you would like to read more about the scientists who contributed to the cell theory, visit *http://www.beyondbooks.com/lif71/4.asp* on the Internet.

Cells come in many sizes and shapes. While most cells are too small to be seen without a microscope, some are quite large. An ostrich egg can be considered a single cell. A single nerve cell can be as long as your leg, but it is much too thin to be visible without a microscope. There are

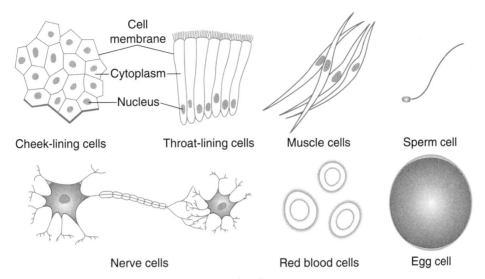

Figure 6.2-1. Several different types of cells found in humans.

many different types of cells, but they all have a similar purpose—to carry out life functions. Figure 6.2-1 illustrates several different types of cells.

The Internal Structure of a Cell

Cells contain structures inside them that help them perform life functions. The most basic of these structures are the nucleus, cytoplasm, and cell membrane. The **nucleus** controls cell activities. The **cell membrane**, the "skin" of the cell, controls the movement of materials into and out of the cell. Between the cell membrane and the nucleus is a thick fluid called **cytoplasm**. Within the cytoplasm are structures, called **organelles**, which are needed to perform life functions. (See Figure 6.2-2 on the next page.) Each organelle is responsible for a specific life function. Table 6.2-1 on the next page lists some of these organelles and their functions. Notice that, except for photosynthesis, these are the *same* life functions you learned about in Lesson 6.1. You will learn more about organelles next year.

According to the cell theory, all organisms are made up of one or more cells. Some organisms, like the ameba, paramecium, and euglena, illustrated in Figure 6.2-3 on page 237, are made up of one single cell. Other organisms, like humans, consist of trillions of cells. An ameba is **unicellular** (*uni* means one), while a human is **multicellular** (*multi* means many); but they are both made up of cells. This is why cells are called the basic units of *structure* of all living things.

Whether a cell is just one of many in an organism, or is the entire organism, it must reproduce, grow, take in food, excrete wastes, and produce energy through respiration. In other words, it must carry out life functions. This is why the cell is considered the basic unit of *function* of living things. The study of cells, called *cytology* (sye-TAHL-uh-gee), explores both the structures and the functions of cells. Cytology is one of the most important branches of biology.

Table 6.2-1. Some Cell Structures and Their Functions

Structure	Function
Mitochondria (my-toh-KAHN-dree-uh)	Respiration—where food is "burned", when food combines with oxygen it produces energy. Called the "power house of the cell"
Ribosomes (RYE-buh-sohms)	Synthesis—where proteins are made
Endoplasmic reticulum (EHN-doh-PLAZ-mihk rhe-TIK-you-luhm)	Transport—move materials within the cell
Nucleus (NEW-klee-us)	Reproduction—where genetic material is stored
Vacuole (VAK-you-ohl)	Digestion and excretion—where digestion occurs or where excess fluid is stored. Vacuoles are usually larger in plants
Chloroplasts (KLOR-oh-plasts)	Photosynthesis—where glucose (sugar) is produced in green plants. Chloroplasts are present in plant cells but not in animal cells

In Lesson 6.1, you learned that humans and plants are similar. They are both considered organisms because they are both made of cells and both perform life functions. The cells in plants and the cells in animals share many of the same structures, but they also have some important differences. (See Figure 6.2-2.) Plant cells

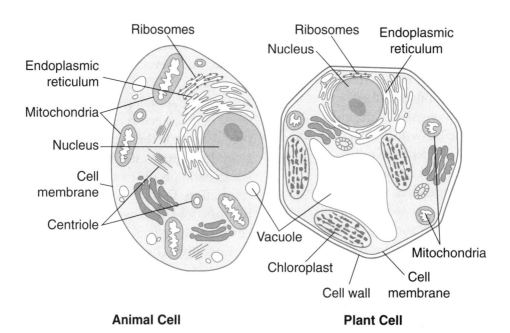

Animal Cell   **Plant Cell**

Figure 6.2-2. The internal structures of a typical animal cell and a typical plant cell.

Figure 6.2-3. Unicellular organisms.

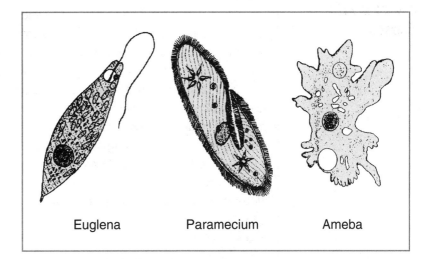

Euglena   Paramecium   Ameba

have a cell wall that encloses the entire cell, including the cell membrane. The tough cell wall gives support to the plant's structure. Animal cells do **not** have cell walls and are enclosed only by the cell membrane.

You may have noticed that there is one function listed in Table 6.2-1 that occurs only in plants. Green plants are able to manufacture their own food through a process called **photosynthesis** (see Lesson 7.8.) Plant cells contain structures called *chloroplasts*. Chloroplasts contain a green chemical called chlorophyll (KLOR-uh-fill), which must be present in order for photosynthesis to take place. Animal cells do not contain chloroplasts; therefore animals cannot manufacture their own food. Animal cells contain structures called *centrioles*, which participate in cell reproduction. Plant cells do not contain centrioles.

Activity

Visit *http://www.teachersdomain.org/6-8/sci/life/stru/index.html* and watch the Single-Celled Organism video by clicking on the *View* button. Then click *Take a Test Drive* and then click the *View* button again. (If you get a window that says "MIME Type Configuration," click NO.) You can Turn Captions *On* to get help with spelling of some new words. Use the information in the video to answer the following questions:

1) What protects the DNA in "eukaryotic" cells?

2) Why is a euglena green?

3) List three materials whose flow is controlled by the cell membrane.

4) Describe how a paramecium gets rid of excess water.

Interesting Facts About Cells

Most of the cells of your body are specialized. They each have a particular function and a special shape and structure to help them perform that function. Nerve cells, for example, specialize in carrying messages. Red blood cells specialize in carrying oxygen, and muscle cells in making things move. There are some cells, however, that do not have a special function. These cells are called stem cells. Stems cells are able to reproduce themselves, providing new stem cells. However, under the right conditions, stem cells can turn into different kinds of specialized cells. Scientists are studying stem cells and trying to find ways to use stem cells to treat disease. Imagine using stem cells to replace the diseased cells of an organ, such as the liver or the pancreas. For more information on stem cell research visit *http://www.sciencenewsforkids.org/ articles/20051019/Feature1.asp* on the Internet.

Questions

1. Which of the following statements **disagrees** with the cell theory?
   (1) All living things are made up of cells.
   (2) Cells are the basic units of structure in all living things.
   (3) Cells are the basic units of function in all living things.
   (4) Cells can be created from nonliving things.

2. The part of an animal cell that controls the flow of materials into and out of the cell is called the
   (1) nucleus   (3) cell membrane
   (2) chloroplast   (4) cell wall

3. Which organelle is found only in animal cells?
   (1) chloroplasts   (3) cell wall
   (2) nucleus   (4) centrioles

4. A cell has often been described as a factory. Which cell structure is correctly matched with an activity that takes place in a factory?
   (1) cell membrane—the powerhouse that produces energy
   (2) ribosome—a machine that makes proteins
   (3) mitochondria—the conveyer belt that moves materials along
   (4) vacuole—the gate that lets things in and out

5. Which term below includes the other three?
   (1) endoplasmic reticulum
   (2) mitochondria
   (3) organelle
   (4) ribosome

6. In which process is oxygen used to release the energy stored in food?
   (1) photosynthesis
   (2) respiration
   (3) digestion
   (4) reproduction

Thinking and Analyzing

1. Identify three structures that are found in both plant and animal cells and describe their functions.

2. Explain why the cell is considered the basic unit of structure and function of living things.

3. The diagram shows a cell with some basic cell structures labeled.

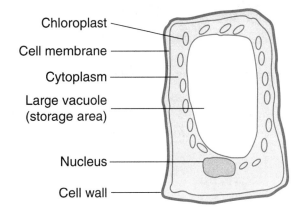

Chloroplast

Cell membrane

Cytoplasm

Large vacuole (storage area)

Nucleus

Cell wall

Identify two structures labeled in the diagram, other than the large vacuole, that indicates this cell is a plant cell.

How Do We Classify Living Things?

Objectives

Identify the five kingdoms of living things.

Classify organisms into the correct kingdom based on cell structure.

Identify examples of organisms belonging to each kingdom.

Terms

kingdom: the most general category of living things

monera (muh-NAIR-uh): a kingdom of unicellular organisms whose cells do not have a nucleus—such as bacteria

protists (PROE-tists): a kingdom of mostly unicellular organisms whose cells have a nucleus

animals: a kingdom of multicellular organisms whose cells contain centrioles

plants: a kingdom of multicellular organisms whose cells contain chloroplasts and cell walls

fungi (FUN-jye): a kingdom of organisms whose cells contain cell walls but no chloroplasts

algae (AL-jee): a one-celled organism that contains chloroplasts and cell walls

protozoa (PRO-tuh-ZOH-uh): a one-celled organism that does not have chloroplasts or cell walls

classification: organization according to similarities in structure

---

Kingdoms

Recall the similarities between you and a green plant. You are both made up of many cells, and you carry out a similar list of life functions. This description of a multicellular organism fits a honeybee as well. Which organism is more similar to a human being, the bee or the plant? How might you decide? If you examined the cells you would see that both the bee and the human are made up of animal cells. Plant cells look very different from animal cells. They have different structures, such as cell walls and chloroplasts. (See Figure 6.2-2 on page 236.) Scientists classify living things on the basis of similarities in structure.

In classifying organisms, scientists first group them into five large categories called **kingdoms.** Organisms in three of these kingdoms are shown in Figure 6.3-1. Lions and honeybees belong to the same kingdom, the animal kingdom. (Scientists

Figure 6.3-1 Examples of animals, plants, and fungi—three of the five kingdoms.

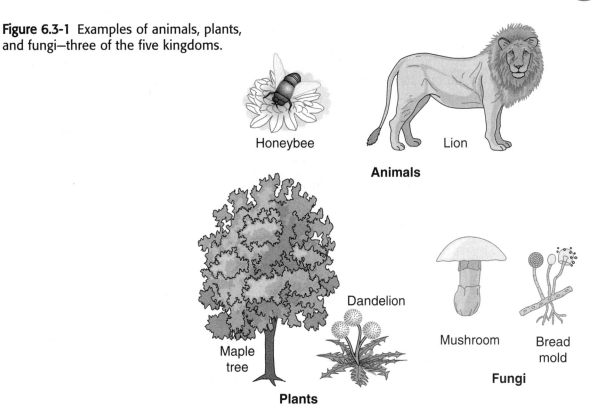

Honeybee

Lion

Animals

Dandelion

Maple tree

Plants

Mushroom

Bread mold

Fungi

often use the Latin names of classification. The Latin names for the kingdoms are listed in Table 6.3-1 on the next page.) Maple trees and dandelions belong to the plant kingdom, while mushrooms and bread mold are classified as fungi. Scientists have identified two additional kingdoms, **monera** and **protists.** While organisms that are in the same kingdom may differ enormously, they are more similar to each other than they are to organisms in the other kingdoms.

Cell Structure and Kingdoms

How do scientists decide in which kingdom an organism belongs? The cell is the basic unit of living things, so we might begin by examining the cells. The **animal** kingdom includes multicellular organisms that have cells containing centrioles but do not have cell walls or chloroplasts. Organisms in the plant kingdom are also multicellular. **Plant** cells have cell walls and chloroplasts. The **fungi** kingdom includes organisms whose cells have cell walls but no chloroplasts. They do not make their own food, as plants do. The monera kingdom includes bacteria whose cells do not contain an organized nucleus.

The protists kingdom is made up mostly of one-celled plantlike and animal-like organisms. Some of these plantlike organisms, called **algae**, contain chloroplasts. The animal-like organisms are called **protozoa** (*proto* means "first" and *zoa* means "animal"). Their cells contain centrioles. The three organisms illustrated in Figure 6.2-3 are examples of protists.

Table 6.3-1. The Five Kingdoms of Living Things

Kingdom	Description	Examples	Latin Name
Animals	Multicellular organisms that have cells that contain centrioles	Insects, fish, birds, mammals, reptiles	Animalia
Plants	Multicellular organisms that have cells enclosed by a cell wall and that contain chloroplasts to make their own food	Trees, shrubs, grasses, mosses	Plantae
Fungi	One-celled and multicellular organisms that have cells enclosed by cell walls, but do not contain chloroplasts	Yeast, molds, mushrooms	Fungi
Protists	Mostly one-celled plantlike and animal-like organisms	Algae, protozoa (e.g., ameba, paramecium)	Protista
Monera	One-celled organisms that lack an organized, membrane-enclosed nucleus	Bacteria, blue-green bacteria (algae)	Monera

Monera have cells that do not contain an organized nucleus. The genetic material is found in the cytoplasm instead of in a nucleus. Such organisms are called *bacteria*. Examples of organisms in each of the five kingdoms are listed in Table 6.3-1.

Beyond Kingdoms

A kingdom may include organisms that differ in many important ways. Birds, cats, frogs, flies, and worms are all animals. You probably realize that cats are more similar to humans than are any of those other animals. Humans and cats both belong to a smaller group of animals called mammals. Scientists first classify organisms into kingdoms, but then they break kingdoms down into smaller and smaller groups, based on similar structures. The classification system used today is very similar to one designed almost 300 years ago.

In 1737, a Swedish scientist named Carolus Linnaeus devised a **classification** system for living things. He grouped them based on internal and external structures and other shared characteristics. Linnaeus called the largest group a *kingdom*, and identified two kingdoms—plants and animals. As you know, scientists now recognize at least five kingdoms of living things.

Linnaeus classified living things into successively smaller groupings called phylum, class, order, family, genus, and species. Humans and cats both belong to the phylum chordate—animals with spinal cords. They also both belong to the class *mammalia*, which means they are mammals. However, humans are in the order primates, while cats are carnivores. Primates are animals that walk standing up on two legs. Primates include the closest relatives of humans, the apes.

As you go from kingdom to species, the groups get smaller, and the organisms in them become more and more similar.

The last two groups are used to give the organism its scientific name. Humans belong to the genus *Homo*, and the species *sapiens*.

Table 6.3-2. Classification of Living Things

Group	House Cat	Red Maple	Lion	Human	Sugar Maple
Kingdom	Animal	Plant	Animal	Animal	Plant
Phylum	Chordate	Tracheophyte	Chordate	Chordate	Tracheophyte
Class	Mammal	Angiosperm	Mammal	Mammal	Angiosperm
Order	Carnivore	Dicotyledon	Carnivore	Primate	Dicotyledon
Family	Felidae	Aceraceae	Felidae	Hominidae	Aceraceae
Genus	*Felis*	*Acer*	*Panthera*	*Homo*	*Acer*
Species	*catus*	*rubrum*	*leo*	*sapiens*	*saccharum*

Stop and Think: Which two organisms in Table 6.3-2 are most closely related? **Answer:** The red maple and the sugar maple have the same genus and therefore are the most closely related.

The scientific term for a human being is *Homo sapiens*. In scientific naming, the genus is always capitalized, and the species is not. Both are written in italics. The scientific name for a wolf is *Canis lupus*. The scientific name for a dog is *Canis familiaris*. When the genus name is the same, it indicates that the organisms are closely related.

Table 6.3-2 lists the classifications of some common organisms. Notice that a housecat and a lion are in the same family, *Felidae,* which means "*catlike.*" They are in a different genus, however. What other animal might be in the same genus as the lion? Tigers, jaguars, and leopards all belong to the genus *Panthera.* (See Figure 6.3-2.)

Figure 6.3-2 The genus *Panthera* includes lions, tigers, jaguars, and leopards.

Jaguar
Panthera onca

Lion
Panthera leo

Leopard
Panthera pardus

Tiger
Panthera tigris

Activity

You are going to classify the food in your pantry. To do this, you are going to identify at least twelve items and classify them into two groups. Then you are going to take one of those groups and classify it into two smaller groups. Explain the reasons for your choices.

For example, you may find the following items in your refrigerator:

Milk, orange juice, soda, hot dogs, bread, eggs, oranges, lemons, grapes, apples, lettuce, cucumber, tomatoes, and carrots.

One possible classification might be to separate the liquids from the solids. The solids can then be broken down into produce (fruits and vegetable) and nonproduce (eggs, hot dogs, and bread). The produce can be broken down into fruits and vegetables. You can further classify your vegetables as green vegetables and nongreen vegetables.

As you can see, you can continue classifying until you are left with small groups of similar items. The important part of this exercise is explaining the reasons for your choices. There are no "wrong" answers.

Interesting Facts About Classification

It is not always clear whether two organisms are the same species. Scientists generally consider organisms to be of one species if they can reproduce offspring that can also reproduce. German shepherds and poodles may look very different, but they are of the same species. They are both dogs, *Canis familiaris*. Wolves belong to a different species, *Canis lupus*. However, dogs and wolves can mate successfully. Some biologists insist that wolves and dogs are not really separate species at all! What do you think?

Questions

1. Which kingdom is correctly paired with its characteristics?
   (1) animals—cell walls and chloroplasts
   (2) plants—centrioles and no cell walls
   (3) protists—cell walls, but no chloroplasts
   (4) monera—no organized nucleus

2. Which kingdoms contain mainly unicellular organisms?
   (1) monera and protists
   (2) plants and fungi
   (3) protists and animals
   (4) plants and animals

3. Phytoplankton are one-celled organisms that live in water. Their cells contain a nucleus and chloroplasts. Phytoplankton are best classified as
   (1) animals   (3) protists
   (2) monera   (4) fungi

4. Which two organisms belong in the **same** kingdom?
   (1) trees and molds
   (2) humans and lobsters
   (3) bacteria and molds
   (4) mushrooms and bees

5. Based on Table 6.3-2, which two organisms are in the same class, but a different order?
   (1) house cat and lion
   (2) red maple and sugar maple
   (3) lion and human
   (4) house cat and sugar maple

Thinking and Analyzing

1. Identify the kingdom for each of the following organisms:

Pine tree   Ameba

Mushroom   Earthworm

You may wish to go to the library or do some research on the Internet to answer this question.

2. The original system of classification, designed by Carolus Linnaeus in the early 1700s, contained only two kingdoms, animals and plants. How did scientists classify protozoa and algae in the 1700s? On what characteristics did they base their decision?

Base your answers to questions 3 and 4 on the table below, which gives the classification of several animals.

3. Select the two animals that are most closely related. Explain how you can tell.

4. Which animal is least like the other four? Explain your answer.

Group	Wolf	Dog	Horse	Grasshopper	Chimpanzee
Kingdom	Animal	Animal	Animal	Animal	Animal
Phylum	Chordate	Chordate	Chordate	Arthropod	Chordate
Class	Mammal	Mammal	Mammal	Insect	Mammal
Order	Carnivore	Carnivore	Ungulate	Orthoptera	Primates
Family	Canidae	Canidae	Equidae	Locustidae	Pongidae
Genus	*Canis*	*Canis*	*Equus*	*Schistocerca*	*Pan*
Species	*lupus*	*familiaris*	*caballus*	*americana*	*troglodytes*

5. Look at the flowchart below which shows how to classify organisms into five kingdoms based on cell structures.

Complete the flow chart by identifying the five kingdoms in the empty spaces labeled 1–5.

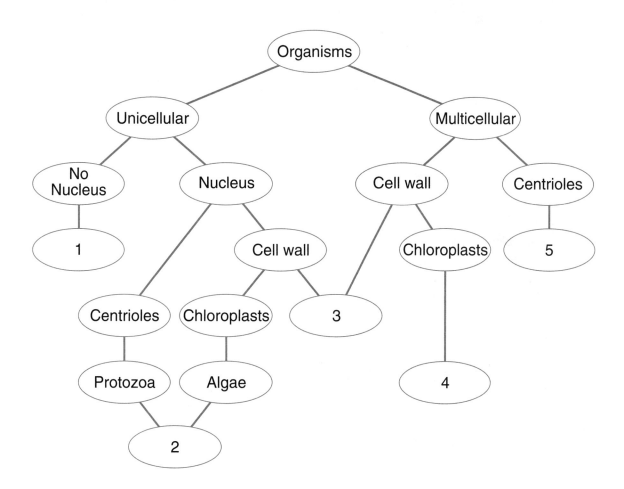

How Have Organisms Adapted for Survival?

Objectives

Distinguish between cold-blooded and warm-blooded animals.

Explain how physical adaptations help organisms regulate body temperature.

Explain how behavioral adaptations help animals survive changes in temperature.

Terms

cold-blooded: an organism that does not maintain a constant body temperature

warm-blooded: an organism that maintains a constant body temperature

adaptation (ah-dap-TAY-shun)**:** a special characteristic that enables an organism to survive under a given set of conditions

physical adaptations: body structures that help an organism survive under a given set of conditions

migration (my-GRAY-shun)**:** seasonal movement of animals from one location to another

behavioral adaptation: activities performed by an organism that help it survive under a given set of conditions

hibernation (HI-buhr-NAY-shun)**:** a sleeplike state of reduced body activity

dormancy (door-MUHN-see)**:** a state in which an organism is inactive while it awaits more favorable environmental conditions

---

How does your body respond to the high temperatures of a hot summer day? You probably sweat, especially if you are active. Sweating is one way that your body maintains a constant temperature of about 37°C (98.6°F). Evaporation (see Lesson 4.5) is a cooling process. As the sweat evaporates, it cools the surface of your skin. You may also have noticed that you feel a chill when you step out of the shower. As water evaporates, it cools your skin.

Cold-blooded Animals

Not all animals need to maintain a constant body temperature. Animals that do *not* maintain a constant body temperature are called **cold-blooded** animals. The body temperatures of cold-blooded animals depend on the temperature of their surroundings. If they cannot obtain enough heat from their surroundings, cold-blooded animals become inactive, or may even die.

Cold-blooded animals

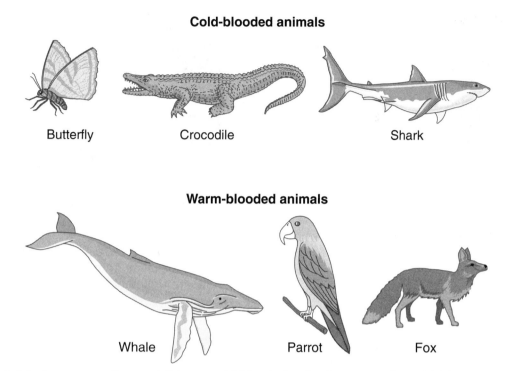

Butterfly   Crocodile   Shark

Warm-blooded animals

Whale   Parrot   Fox

Figure 6.4-1. Insects, reptiles, and fish are cold-blooded animals. Mammals and birds are warm-blooded animals.

Fish and reptiles are cold blooded. If you have a fish tank at home, you know that you must maintain a certain water temperature to keep the fish alive. A heat lamp, or heated rock, is often placed in a tank or cage to provide warmth for a pet snake. (See Figure 6.4-1.)

Warm-blooded Animals

Have you ever had a cat sit on your lap? You may have noticed how warm it feels. Like humans, cats are **warm-blooded** animals. Warm-blooded animals maintain a constant body temperature no matter what the surrounding temperature may be. Cats maintain a temperature of about 39°C (101°F.) Since this is higher than your body temperature of 37°C, the cat feels warm to

you. Cats are *mammals*. Mammals are animals with hair or fur and supply milk for their young. Of all the different classes of animals, only birds and mammals are warm-blooded.

As you learned in Lesson 6.1, respiration occurs in the cells of organisms to release energy stored in food. Warm-blooded animals, like humans, use some of this energy to maintain a constant body temperature. If we are unable to maintain our body temperature, we cannot survive. We may respond to the cold by shivering, putting on a coat, or even getting "goose bumps." Other warm-blooded animals may respond in different ways. Organisms have special characteristics, called **adaptations**, that enable them to survive under a given set of conditions.

Physical Adaptations

Organisms have developed certain structures that help them survive under a given set of conditions. These structures are called **physical adaptations**. Whales, for example, develop a thick layer of fat called blubber under their skin. The blubber protects the whales from extreme cold. Many mammals develop thick coats of fur to keep them warm in winter. In the spring some mammals, like many types of dogs, shed their winter coat and replace it with a shorter, lighter one.

Animals also need to protect themselves from overheating. As you know, humans sweat to help them cool down. Dogs, however, have developed a different way of staying cool. When a dog pants, it is trying to cool off by losing heat from its tongue. Foxes lose heat through their ears. As a result, foxes living in cool areas usually have smaller ears than foxes living in warmer areas. More heat radiates from the larger ears, keeping the fox cool.

Behavioral Adaptations

Many birds respond to seasonal changes in temperature by moving to a new location. Snow geese fly south for most of the winter, returning in early March. (See Figure 6.4-2.) This behavior is called **migration**. Migration is an example of a **behavioral adaptation**, a special behavior that helps an organism adapt to changes in its environment.

Some organisms, such as bears and chipmunks, fall into a deep sleep for the entire winter. This behavior is called **hibernation**. During hibernation, body activities slow down and body temperature

Figure 6.4-2. Snow geese pass through New York when migrating from northern Canada to the southern United States for the winter.

drops. A hibernating animal requires less energy to keep it alive. This allows the animal to survive the entire winter on body fat that it stored in the fall.

Dormancy

Other living things may adjust to extreme environmental changes by entering a state of **dormancy**, becoming completely inactive. During the winter, when a tree has lost its leaves, the tree is not dead—it is *dormant*. When spring comes, bringing warmer conditions, the tree grows new leaves.

Some plants die when the weather gets cold. However, they produce seeds, which may remain dormant through the winter. When the soil warms in the spring, the seeds produce new plants. To learn more about how animals spend the winter, visit *http://www.sciencemadesimple.com/animals .html* on the Internet.

Activity

Visit *http://www.teachersdomain.org/6-8/sci/life/reg/index.html*
and watch the Fever video by clicking on the *View* button. Then
click *Take a Test Drive,* and then click the *View* button again. (If
you get a window that says "MIME Type Configuration," click
NO.) You can Turn Captions *On* to get help with spelling of
some new words. Use the information in the video to answer the
following questions:

1. What is homeostasis?

2. What part of the brain controls your body's set point?

3. How does stress affect body temperature?

4. What is the purpose of fever?

Interesting Facts About Keeping Warm

When animals are cold, they may respond by fluffing up their fur or hair. Small muscles in the skin allow the hair to stand up. The hair helps protect the animal from the cold. Humans do not have much hair, but the small muscles in the skin are still active. They can still make their hair "stand up." We experience "goose bumps" when the skin responds in this way. Only mammals have hair or fur, so only mammals can get goose bumps. A goose cannot.

Questions

1. One of the main purposes of sweating is to
   (1) cool the body
   (2) keep the skin dry
   (3) warm the body
   (4) keep the skin moist

2. Which of the following animals is cold-blooded?
   (1) a fox   (3) a sparrow
   (2) a snake   (4) a cat

3. Fur is a
   (1) behavioral adaptation to prevent a loss of heat
   (2) behavioral adaptation to cause a loss of heat
   (3) physical adaptation to prevent a loss of heat
   (4) physical adaptation to cause a loss of heat

4. Some insect-eating birds migrate to warmer areas in the winter due to a shortage of food. Which of the following would best explain their migration?
   (1) insects are warm-blooded
   (2) birds are warm-blooded
   (3) insects are cold-blooded
   (4) birds are cold-blooded

Thinking and Analyzing

1. Why is seeing a robin considered the "first sign of spring?"

Base your answers to questions 2 through 4 on the information below and on your knowledge of science.

   A warm-blooded animal maintains a constant body temperature no matter what the temperature of its surroundings may be. This ability is important to the organism's survival. These organisms have many different body structures and behaviors that help maintain a constant body temperature.

2. Whales have a thick layer of blubber (fat) under their skin. How does this blubber help the whales to maintain a constant body temperature?

3. Humans sweat when they are in the hot sun. How does sweating help humans to maintain a constant body temperature?

4. Foxes living in different parts of the world have different-sized ears. The arctic fox, which lives in cold climates, has small ears. The desert fox, which lives in hot climates, has large ears. How does ear size help each of these foxes maintain a constant body temperature?

6.5 How Do Organisms Move?

Objectives

Identify reasons why organisms need to move from place to place.

Identify three different types of locomotion in unicellular organisms.

Terms

flagellum (fluh-JELL-uhm): a tail-like structure used for locomotion

cilia (SIH-lee-uh): tiny hairlike structures used for locomotion

pseudopod (SOO-doh-pod): an extension of cytoplasm used by an ameba for locomotion

Locomotion in Animals

Multicellular animals need to move from place to place. We call this life function *locomotion*. In the previous lesson, you learned that moving from place to place is one important way that organisms adapt to change. Other reasons for moving from place to place include finding a mate, finding food, and escaping danger. Animals can do this by swimming, flying, or walking. For this purpose, animals have developed physical adaptations such as fins for swimming, wings for flying, and legs for walking.

We generally expect fish to swim, birds to fly, and mammals to walk. This is not always the case. (See Figure 6.5-1.) Marine mammals, such as whales and dolphins, spend their entire lives in the ocean. These organisms have developed fins and tails that allow them to swim. Penguins are birds, but

they cannot fly. Their wings are adapted for swimming rather than flying. The ostrich, another flightless bird, has long legs adapted for running. Bats are mammals that are able to fly.

Locomotion in Protozoa

Protozoa also need to move from place to place for many of the same reasons. How can one-celled protozoa (see Figure 6.2-3 on page 237) accomplish this? The protozoa have organelles that are specifically designed to aid in locomotion.

The three organisms shown in Figure 6.2-3 use three different methods of locomotion. The euglena has a long tail-like structure called a **flagellum**. This tail spins around and acts like a propeller to move the organism. If you look at the photograph of the sperm cells and egg cell at the beginning of the chapter, you

Penguin   Ostrich   Bat

Whale   Dolphin

Figure 6.5-1. Adaptation for locomotion—flightless birds and mammals that swim or fly.

will notice that one human cell, a sperm cell, has a flagellum as well.

The paramecium shown in Figure 6.2-3 is surrounded by tiny hairlike structures called **cilia**. These cilia move together, causing the paramecium to swim.

The ameba does not have a special structure to help it move. Instead, the flexible cell membrane of the ameba extends out, forming leglike structures called **pseudopods** (*pseudo* means fake, and *pod* means foot). The movement of these pseudopods enables the ameba to creep along.

Activity

Revisit *http://www.teachersdomain.org/6-8/sci/life/stru/index .html* and watch the Single-Celled Organism video by clicking on the *View* button. Then click *Take a Test Drive* and then click the *View* button again. (If you get a window that says "MIME Type Configuration;" click NO.) You can Turn Captions *On* to get help with spelling of some new words.

Observe the three methods of locomotion shown in the video.

1. List three methods of locomotion in protozoa.

2. Describe the structures that are used in each of these three methods of locomotion.

Interesting Facts About Locomotion

You may have seen a snake move in a zig-zag motion. This motion is the fastest way for a snake to move. This motion is also used for swimming. Snakes move in other ways as well. Some snakes creep, jump, or move sideways or diagonally (see the figure below). The most unusual method of locomotion in a snake, however, is flying. The flying snake coils itself up and stretches out quickly and jumps from the limb of a tree. Its flattened body acts like a parachute. It can glide up to 100 meters (over 300 feet). With all of these different adaptations for locomotion, it may surprise you to find out that snakes cannot move backward.

A sidewinder is a type of rattlesnake that moves by thrusting its body diagonally forward in a series of flat S-shaped curves.

Questions

1. Which animal is properly paired with its structure used for locomotion?
   (1) ostrich—wings
   (2) ameba—flagellum
   (3) paramecium—cilia
   (4) dolphin—legs

2. Which of the following is **not** a good reason for an organism to move from place to place?
   (1) to find a mate   (3) to find danger
   (2) to find food   (4) to find shelter

3. All of the following structures are involved in locomotion **except**
   (1) the tail of a whale
   (2) the roots of a maple tree
   (3) the wings of a penguin
   (4) the fins on a dolphin

4. The purpose of cilia or a flagellum in a unicellular organism is most closely related to the life function
   (1) nutrition   (3) transport
   (2) reproduction   (4) locomotion

5. Look at the photo of the sperm cells at the beginning of this chapter. One sperm cell (the smaller cell) combines with the egg cell (the larger cell) during the process of reproduction. What method of locomotion does a sperm cell use to reach the egg?
   (1) cilia   (3) flagellum
   (2) pseudopod   (4) walking

Thinking and Analyzing

1. A unicellular organism that contains cilia is called a ciliate. A unicellular organism that contains a flagellum is called a flagellate. Compare and contrast how ciliates and flagellates move.

2. State three reasons why animals need to move from place to place.

Review Questions

Term Identification

Each question below shows two terms from Chapter 6. One of the terms is defined.
(1) Choose the term that matches the definition.
(2) Describe how the two terms are different.
Following each term is the section (in parenthesis) where the description or definition of that term is found.

1. *Organelle (6.2) — Cytoplasm (6.2)*
   A small structure within the cell that carries out a specific life function

2. *Multicellular (6.2) — Unicellular (6.2)*
   An organism that consists of only a single cell

3. *Cell membrane (6.2) — Nucleus (6.2)*
   A structure that surrounds the cell and controls the movement of materials into and out of the cell.

4. *Protist (6.3) — Monera (6.3)*
   A kingdom of unicellular organisms whose cells do not have a nucleus

5. *Plants (6.3) — Fungi (6.3)*
   A kingdom of multicellular organisms whose cells contain cell walls but no chloroplasts

6. *Flagellum (6.5) — Cilia (6.5)*
   A tail-like structure used for locomotion

Multiple Choice (Part 1)

Choose the response that best completes the sentence or answers the question.

1. Which of the organisms shown below consists of only one cell?

Pine tree   Ameba   Mushroom   Earthworm

   (1) pine tree   (3) mushroom
   (2) ameba   (4) earthworm

2. Which structures are found in **all** living things?
   (1) cells   (3) chloroplasts
   (2) lungs   (4) cilia

3. Which of the following organisms is multicellular?
   (1) flea   (3) euglena
   (2) paramecium   (4) ameba

4. The purpose of the flagellum on a unicellular organism is best described as
   (1) respiration   (3) photosynthesis
   (2) locomotion   (4) excretion

5. Both plant cells and animal cells contain
   (1) chloroplasts   (3) mitochondria
   (2) cell walls   (4) centrioles

6. A scientist looking at some tiny organisms through a microscope notices that their cells lack an organized nucleus. The scientist might be looking at
   (1) bacteria   (3) plants
   (2) ameba   (4) paramecia

7. Which of the following is a **physical** adaptation to cold weather?
   (1) flying south
   (2) developing a thick coat of fur
   (3) hibernation
   (4) shivering

8. The purpose of the layer of fat, called blubber, in seals is most similar to the purpose of
   (1) cilia in paramecia
   (2) fur on a mink
   (3) wings in birds
   (4) sweat glands in humans

9. Trees that have lost their leaves in the winter are said to be
   (1) dormant   (3) hibernating
   (2) migrating   (4) dying

10. Which of the following animals is warm-blooded?
   (1) a turtle   (3) an alligator
   (2) an ostrich   (4) a shark

11. After examining the cells of an unknown organism, a scientist concludes that the organism can manufacture its own food. The scientist noticed that the cells contain
   (1) nuclei   (3) chloroplasts
   (2) centrioles   (4) cell walls

Thinking and Analyzing (Part 2)

1. The table below provides some information about common plant cell structures and their functions. In the table, there are three blank spaces. Identify and write the name of the plant cell structure that performs the function described for it.

2. Which organism do you think is more similar to a human— a tree, or a paramecium?
(a) State at least one way in which a human is more similar to a paramecium than to a tree.
(b) State at least one way in which a human is more similar to a tree than to a paramecium.

Base your answers to questions 3 and 4 on the diagram of a cell on the right and on your knowledge of science. Some of the cell structures are labeled in the key.

3. Which *two* structures are found in plant cells, but **not** in animal cells?

4. Select *three* of the structures named in the diagram key. For each structure selected, state the name of the structure and its function in the cell.

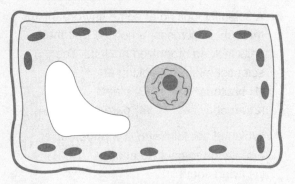

Plant cell

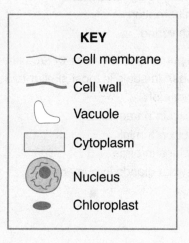

Plant Cell Structure	Function
Cell membrane	Allows substances to enter and leave the cell
	Directs the cell's activities, including reproduction
	Captures energy from sunlight to make food
	Protects and supports the cells
Cytoplasm	Allows the movement of materials around the cell and supports other cell structures
Vacuole	Stores food, water, and waste

5. Have you ever held a snake? Garter snakes are small reptiles that some people keep as pets. Snakes are cold-blooded animals, but if you pick up a pet garter snake you will notice that it does not feel cold. (**You should never pick up a wild snake**.) Warm-blooded animals feel warm to the touch, so why doesn't the cold-blooded snake feel cold?

Read the paragraph below and use the information to answer questions 6 and 7.

A deep-sea explorer discovers a new species. It is multicellular. Its cells do have a nucleus, but do not have chloroplasts or a cell wall.

6. In what kingdom would you classify this organism?

7. What would scientists look at to classify it further?

Chapter Puzzle (*Hint:* The words in this puzzle are terms used in the chapter.)

Across

1 a state in which an organism is inactive while it awaits more favorable conditions

3 an extension of cytoplasm used by an ameba for locomotion

7 tiny hairlike structures used for locomotion

9 ameba and paramecia belong to the kingdom called the _____

10 a special characteristic that enables an organism to survive under a given set of conditions

12 a kingdom of organism whose cells contain cell walls but no chloroplasts

13 seasonal movement of animals from one location to another

14 processes that occur in all living things are called life____

16 the building block of living things

19 unicellular organisms whose cells do not have a nucleus are found in the kingdom ___

20 biology is the study of ___

22 containing more than one cell

23 the structure within the cell that controls cell activities

24 the outer covering of an animal cell is called the cell ___

25 the cell is filled with a gel-like substance called the ___

Down

2 organization according to similarities

4 the process through which plants use the energy in sunlight to produce food

5 a sleeplike state of reduced body activity

6 an animal that does not maintain a constant body temperature is called a ___ blooded animal

8 a small structure within the cytoplasm that carries out a specific life function

10 a kingdom of multicellular organisms whose cells contain centrioles

11 a living thing

14 a tail-like structure used for locomotion

15 consisting of a single cell

17 an animal that maintains a constant temperature is called a ___ blooded animal

18 ___ cells contain a cell wall and chloroplasts

21 the most general category of living things

Chapter 7

Food Chains and Food Webs

Contents

This photo illustrates a predator-prey relationship. The lion, the predator, is about to attack the zebra, its prey.

What Is This Chapter About?

Ecology is the study of the way living things interact with their environment, or surroundings. The environment includes both living and nonliving things. Living things depend on the environment for food, water, and oxygen.

In this chapter you will learn:

1. Living things require a constant supply of energy.

2. Living things obtain food, water, and oxygen from their environment.

3. The sun is the source of all energy in most environments.

4. Plants use the sun's energy to convert carbon dioxide and water into glucose, a form of sugar.

5. Each animal has a particular role to play in the environment.

6. Many materials are used again, or recycled, in the environment.

Career Planning: Ecologist

Ecologists are scientists who study the relationships between living things and their environment. Some ecologists study how changes in climate affect the environment. Others try to predict how human activities affect organisms. Rachel Carson was a scientist who studied the effects of pesticides on the food chain. Her book, *The Silent Spring*, warned that the chemicals used to kill insects would eventually work their way up the food chain and poison many species of birds. Her work led to a ban on the use of the pesticide DDT.

Internet Sites:

http://www.kidsplanet.org/ A fun Web site where you can play Wild Games or read about the Web of Life.

http://arcytech.org/java/population/facts_foodchain.html Read interesting facts about food chains and food webs with examples and definitions.

http://www.harcourtschool.com/activity/food/food_menu.html Play a game where you drag and drop pictures of animals to make a food web.

http://www.kaweahoaks.com/html/balance_of_nature.html This Web site contains more advanced materials on food webs, food chains, and the balance of nature.

7.1 How Do Living Things Depend on Their Environment?

Objective
Identify the parts of the environment.

Terms
environment (en-VY-run-ment): the surroundings in which an organism lives, and includes both living and nonliving things
ecology: the study of the interaction between organisms and their environment

Interacting With the Environment

Review the life processes listed in Table 6.1-1 on page 230. Although all living things carry out these processes, different types of organisms do so in different ways. For example, nutrition differs in animals and plants. Animals take in food by eating living things or the remains of dead animals. Green plants, on the other hand, are able to manufacture their own food from nonliving things, such as water, carbon dioxide, and sunlight.

Some of these life functions involve an exchange of material between the organism and its surroundings. These surroundings are called the **environment**. The environment includes both living and nonliving things. The nonliving part of an environment includes air, water, soil, light, and temperature. **Ecology** is the branch of science that studies the relationship between living things and their environment.

In the process of *nutrition*, animals eat and digest the food they find in their environment. In *respiration*, animals use oxygen from the environment to burn the sugars they get from the food they eat. In this process, carbon dioxide and water are passed from the animal to the environment. During *excretion*, wastes pass from the animal to the environment. In each case, the organism must interact with both its living and nonliving environment.

Questions

1. An environment consists of
   (1) only living things
   (2) only nonliving things
   (3) both living and nonliving things
   (4) neither living nor nonliving things

2. Which statement is associated with the process of excretion?
   (1) The organism returns waste to the environment.
   (2) The organism removes oxygen from the environment.
   (3) The organism takes in food from the environment.
   (4) The organism takes in carbon dioxide from the environment.

3. Which pair consists of one living and one nonliving thing in an environment?
   (1) plants and animals
   (2) temperature and sunlight
   (3) trees and water
   (4) water and oxygen

Thinking and Analyzing

1. The illustration to the right shows a desert environment. Identify three nonliving things and three living things in this environment.

7.2 How Does the Environment Cycle Materials?

Objectives

Describe how materials such as water, oxygen, carbon dioxide, and nutrients are cycled through the environment.

Describe how living and nonliving things interact in an ecosystem.

Terms

ecosystem: all the living and nonliving things that interact within a certain area

cycle: a periodically repeated sequence of events

water cycle: the process by which water moves back and forth between Earth's surface and the atmosphere

carbon cycle: the process through which plants and animals exchange carbon in the form of sugar and carbon dioxide

Ecosystems

Living things, or organisms, interact with their environment. Organisms obtain food, water, and oxygen from the environment. They release wastes, such as carbon dioxide, back into the environment. Organisms depend on the environment, and the environment depends on the organisms within it. The system of interactions between organisms and their environment in a given area is called an **ecosystem**. An ecosystem may exist in a forest, a desert, a pond, an ocean, or even in a drop of water. Figure 7.2-1 shows an aquarium that contains plants, fish, snails, water, an air pump, gravel, and a heater. This is an example of an ecosystem.

The nonliving things in an ecosystem are essential to the survival of the organisms that live there. All living things require water to survive. Without water, the plants, fish, and

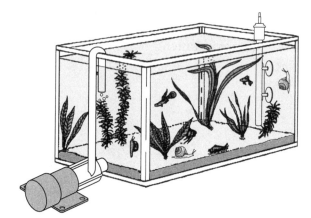

Figure 7.2-1   An aquarium is an example of a small ecosystem.

snails would all die. You, also, could not survive without water. Water is an example of a nonliving thing that living things depend on. A cell, the basic unit of living things, consists mainly of water. Our blood is mostly water as well.

Materials Cycle Through the Environment

The Water Cycle

Organisms obtain water from their environment in many different ways. We obtain water by drinking it and from the foods we eat. Some organisms can absorb water directly into their cells. Most organisms return water to the environment through excretion. (See Lesson 7.1.) We say that water is cycled through the environment. A **cycle** is a series of events that is repeated over and over again. In a cycle, one event leads to another and eventually leads back to the beginning to start over again. In this example, the organism takes water from the environment by drinking it, and then returns the water to the environment by excreting it. Another way that water is cycled through the environment is through rain. Water evaporates (turns to gas), condenses (turns back to liquid), and finally precipitates (falls back to earth as rain) as shown in Figure 7.2-2. We talked about *precipitation* (pree-sihp-uh-TAY-shuhn) in Lesson 5.5. This movement of water back and forth between Earth's surface and the atmosphere is called the **water cycle**.

The Carbon Cycle

Most organisms need oxygen to survive. Animals remove oxygen from the air when they breathe. This oxygen combines with the sugar in the food they eat to produce energy. As we learned in the previous lesson, this process is called *respiration*. During respiration, animals release carbon dioxide, a gas, into the environment. Plants use this carbon dioxide to make food. (We will discuss in detail how plants make food later in this chapter.) When plants make food, they release oxygen into the environment. This process, illustrated in Figure 7.2-3, is called the **carbon cycle**. The carbon cycle is another important cycle in the environment.

Materials Recycle Through the Environment

Over time, materials such as oxygen, water, and nitrogen (see Lesson 7.5) are transferred

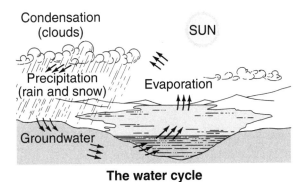

The water cycle

Figure 7.2-2. Water is constantly cycled through the environment.

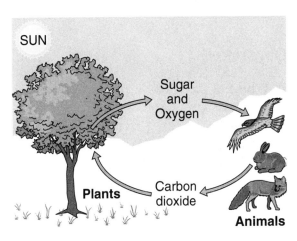

The carbon cycle

Figure 7.2-3. Oxygen and carbon dioxide are constantly cycled in an ecosystem.

from one organism to another and between organisms and their environment. When these materials are used over and over again, they are being *recycled* in the environment. One thing that is not recycled, however, is energy. Energy flows in one direction in an ecosystem.

Energy Is Always Needed

Energy, as you learned in Chapter 2, is defined as the ability to do work. Organisms need energy to perform their life functions.

This energy is usually provided by sunlight. Light and heat are forms of energy. The energy in sunlight is absorbed by green plants and used to produce sugar, a nutrient that stores energy. (See Lesson 7.8.) Sugar provides energy for the plant and for any animal that eats the plant. When an animal eats another animal, the energy stored in the animal that is eaten is *transferred* to the animal that eats it. In this case, energy is *not* returned to the environment. The environment needs a constant supply of energy.

Activity

You are given a large glass dome and are told to fill it with whatever you need to start an environment that will survive on the moon. However, you are limited to ten things.

1. What ten things would you place in your environment?
2. List two things that you will not have to replace in your environment once it is formed and explain why.
3. What must be continuously supplied to your environment? Explain your answer.

Questions

1. Which of the following is **not** recycled in an ecosystem?
   (1) water   (3) carbon dioxide
   (2) oxygen   (4) sunlight

2. The water cycle includes all of the following **except**
   (1) evaporation   (3) condensation
   (2) elimination   (4) precipitation

3. What do all organisms need to survive?
   (1) soil   (3) carbon dioxide
   (2) blood   (4) energy

4. Which of the following directly provides animals with their main source of energy?
   (1) food   (3) carbon dioxide
   (2) water   (4) sunlight

Thinking and Analyzing

Use the diagram below to answer questions 1–4.

1. What is the primary source of energy for this ecosystem?

2. What is the main source of oxygen in this ecosystem?

3. If light is eliminated from this environment, the fish will eventually die. Explain why.

4. Name two processes that combine to maintain the water level in the pond.

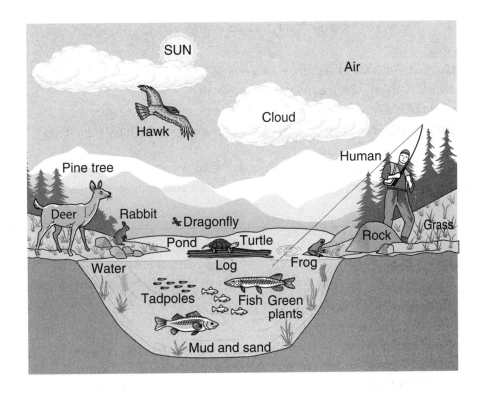

7.3 How Does Energy Flow Within an Environment?

Objective
Describe the ways that organisms get their energy.

Terms
herbivore (UR-bih-vore): a plant-eating animal

carnivore (KAR-nih-vore): a meat-eating animal

omnivore (AHM-nih-vore): an animal that eats both plants and animals

predator: (PREH-dih-tuhr): an animal, such as the lion, that must hunt and kill for food

prey: an animal the predator (for example, the lion) hunts, such as the zebra

scavenger (SKAH-ven-juhr): an animal, such as a vulture, that eats the remains of dead animals

Energy Is Stored in Food

All organisms need energy to survive. Humans get this energy from the nutrients they eat in food. Energy is stored in these nutrients. Sugar is one nutrient that supplies energy. During the digestive process, our bodies change food into a form that can be used to make energy.

Green plants make sugar using sunlight during a process called photosynthesis (*photo* means light, to *synthesize* means to make). During photosynthesis, green plants use the sun's energy to produce sugar from carbon dioxide and water. This sugar provides energy for the plant to live. The plant also absorbs other materials from the soil. These materials, along with the sugar it produces, provide the plant with all the ingredients it needs to survive. We will talk more about photosynthesis in Lesson 7.8.

Animals Get Food in Different Ways

Animals get their energy by eating plants or other animals. Animals that eat only plants are called **herbivores**. A zebra is an example of a herbivore. Animals that eat only other animals are called **carnivores**. A lion is an example of a carnivore. When a lion eats a zebra, it is getting energy indirectly from plants. Energy has passed from the plant, to the zebra that eats the plant, and, finally, to the lion that eats the zebra. Some animals eat both plants and animals. These animals are called **omnivores**. Humans are omnivores, because we eat both plants and animals.

A lion hunts and eats a zebra for its nutrition. (See the photo at the beginning of this chapter.) An animal that hunts other animals is called a **predator**. The animal that it hunts is called its **prey**. Other carnivores feed on the remains of dead animals. These carnivores are called **scavengers**. Hyenas, vultures, and many species of insects are scavengers. In Figure 7.3-1 the vultures wait for the lion to finish eating its kill.

Figure 7.3-1. Vultures wait for the lion to finish eating its kill.

Interesting Facts About Carnivores

Dogs: Carnivores or Omnivores?

We know that a dog loves to eat meat. Nothing makes a dog happier than some steak on a bone. However, some people feed their pet dogs carrots or other vegetables. Many dog foods contain vegetable products. Some dogs have even been raised as vegetarians. So obviously, dogs can eat vegetables. Does that make a dog an omnivore?

The answer is no. Dogs (canines) are carnivores. A dog's digestive system is designed to digest meat. Dogs have sharp teeth for tearing, not flat teeth for grinding. In the wild, dogs will only eat meat. However, dogs are not *obligate* carnivores.

Obligate (AHB-lih-gate), or true, carnivores are animals that can live only on a diet of meat. They may consume cheese, honey, or other foods containing sugar, but do not need these foods to survive. True carnivores lack the ability to digest vegetable matter. Unlike dogs, cats (felines) *are* obligate carnivores and should be fed only meat products. Since many dog foods contain plant products, you should not feed dog food to a cat.

Activity

Read the paragraphs below to answer the questions in this activity.

Grizzly bears are often thought of as ferocious hunters. In reality, most of their diet is made up of grasses, roots, and berries. Occasionally, bears will tear open logs and feast on the ants living inside. Grizzly bears are very good at fishing and often hunt for salmon.

A bear catching salmon.

Salmon are very interesting fish. They spend most of their lives in the oceans. When it is time to reproduce, they return to the rivers and streams where they were born. Young salmon eat small insects, while adult salmon eat small fish and tiny, pink, shrimplike animals called krill . . . lots and lots of krill. The krill give salmon their pink color.

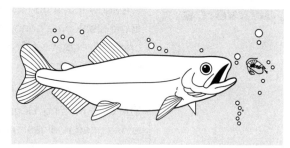

A salmon chasing krill.

Krill are very small. They survive on tiny microscopic plants called phytoplankton (FIE-toe-PLANK-tuhn). These one-celled plants float near the surface of the ocean.

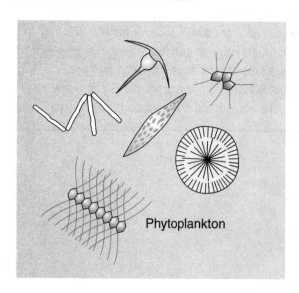

Phytoplankton

Examples of phytoplankton.

1. Salmon are both predator and prey. Explain.

2. Complete the following table. Place a check in the correct box to indicate if the animal is a carnivore, herbivore, or omnivore.

	Carnivore	Herbivore	Omnivore
Bear			
Salmon			
Krill			

3. The grizzly bear is a type of brown bear. Use the Internet to learn more about brown bears and to answer the next three questions. One good site is

http://thebigzoo.com/Animals/Grizzly_Bear.asp

(a) What foods do grizzly bears eat the most?

(b) What provides the grizzly bear with energy during the winter months?

(c) What is the biggest threat to the survival of the grizzly bear?

Questions

1. An animal that hunts other animals for its food is called a
   (1) producer   (3) predator
   (2) herbivore   (4) prey

2. An animal that eats only plants is called a(n)
   (1) herbivore   (3) omnivore
   (2) carnivore   (4) predator

3. Most humans are
   (1) herbivores   (3) omnivores
   (2) carnivores   (4) scavengers

4. Some spiders build webs to catch and eat insects. The tarantula is a type of spider that does not build webs. It hunts other spiders, insects, and even small lizards for its food. Tarantulas are best described as
   (1) herbivores and predators
   (2) herbivores and prey
   (3) carnivores and predators
   (4) carnivores and prey

Thinking and Analyzing

1. What is the original source of energy for an ecosystem? Describe how this energy is transferred to other animals.

2. Explain why predators can never be herbivores.

3. Using the Internet or the library, find out what is meant by "bird of prey." Give two examples.

7.4

What Is a Food Chain?

Objectives
Use a food chain to identify producers, consumers, and decomposers.

Trace the flow of energy through a food chain.

Terms
producer: an organism that makes its own food; most producers are green plants

consumer (kuhn-SUE-muhr): an organism that obtains nutrients by eating other organisms

food chain: a sequence of organisms through which energy is passed along in an ecosystem

decomposer (dee-kuhm-POH-zuhr): special organisms, such as fungi (mushrooms and molds) and some bacteria, that break down dead animals' remains and return their nutrients to the soil

Producers and Consumers

Whether they are herbivores or carnivores, predators or prey, all animals depend on green plants for energy. Green plants are called **producers** because they produce their own food using the sun's energy. Animals are called **consumers** because they eat, or consume, plants and other animals for their food. Every animal depends directly or indirectly on green plants for energy.

The nutrients in green plants are passed along from one organism to another in a sequence called a **food chain**. During photosynthesis, grass produces food. A zebra eats the grass to get its nutrients. A lion, in turn, eats the zebra. We can use an arrow to show how the nutrients, or energy, flow from organism to organism.

Grass → Zebra → Lion

Decomposers

When an organism dies, it decays. Special organisms, called **decomposers**, break down the remains and return some of its nutrients to the soil. Plants, such as grass, can then use these nutrients again. Nitrogen is one of the important nutrients that is cycled through the ecosystem by the decomposers.

Decomposers include fungi, such as mushrooms and molds, and some bacteria. Fungi and bacteria cannot make their own food, so they feed on living or dead organisms. Decomposers are considered the last link in any food chain. Figure 7.4-1 shows an example of a food chain.

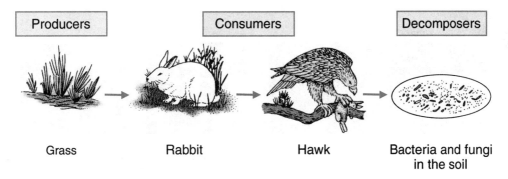

| Producers | Consumers | Decomposers |
| Grass | Rabbit | Hawk | Bacteria and fungi in the soil |

Figure 7.4-1. A food chain.

Activity

Use the Internet and go to *http://www.kidsplanet.org/* and do the Web of Life activity. Draw a food chain based on some of the animals in the activity. (Make sure that you draw the arrows in the direction of the energy flow.)

Questions

Use Figure 7.4-1 to answer questions 1–3.

1. Which organism is the producer in this food chain?
   (1) grass   (3) hawk
   (2) rabbit   (4) bacteria and fungi

2. In this food chain, the hawk is a
   (1) producer   (3) decomposer
   (2) consumer   (4) herbivore

3. Which organism is a herbivore?
   (1) grass   (3) hawk
   (2) rabbit   (4) bacteria and fungi

Thinking and Analyzing

Base your answers to questions 1–3 on the figure on page 269 and your knowledge of science. The diagram represents woodland and pond ecosystems containing a variety of plants and animals.

1. Describe how the sun's energy is transferred to the animals in this ecosystem.

2. Identify one consumer in this ecosystem.

3. In this ecosystem, the rabbits eat the grass and the hawks eat the young rabbits.

Copy the table below into your notebook and identify the role—producer, consumer, or decomposer—each organism plays in this food chain.

Organism	Role
Grass	
Rabbit	
Hawk	

4. What is the role of fungi and bacteria in the food chain?

7.5 How Do Food Chains Interact?

Objectives

Interpret a food web.

Identify the various food chains within a food web.

Identify the role of each organism in the food web.

Term

food web: a number of interconnected food chains

Food Webs

Figure 7.4-1 on page 275 shows a food chain, in which hawks eat rabbits. Suppose a disease causes all of the rabbits to die. What would happen to the hawks? If the hawks did not have another food source, they would die as well. Hawks also eat mice, chipmunks, and other birds. An environment contains many overlapping food chains because most animals eat more than one kind of food. A diagram that shows several connected food chains is called a **food web**. Figure 7.5-1 shows a simple food web.

Notice how the food web contains many food chains. One of these, grass → rabbits → hawks → bacteria and fungi, is the same as the food chain in Figure 7.4-1. How many other food chains can you find in this food web?

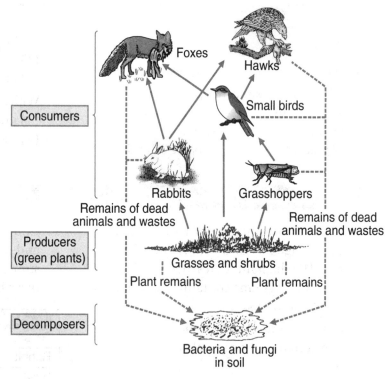

Figure 7.5-1. A food web consists of several interconnected food chains. There are six food chains in this food web. Can you find them?

Direction of Energy Flow

Figure 7.5-2 below shows another food web. Notice how the arrows in the food web always go *from* the food *to* the animal that eats that food, because that is the way the energy flows. If there are predators in the food web, the arrow always points *from* the prey *to* the predator. **Stop and Think:** There

are three producers in this food web. Can you find them? **Answer:** carrots, grass, and grain.

This particular food web leaves out the decomposers. If the decomposers were included, there would have to be an arrow leading *from* every one of the nine organisms shown, *to* the decomposers.

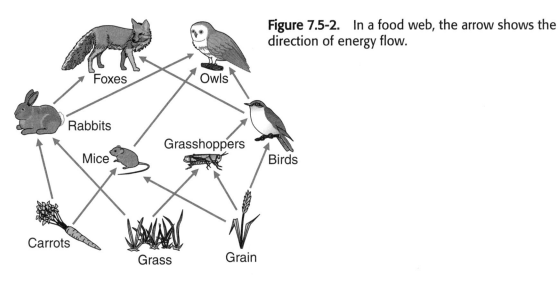

Figure 7.5-2. In a food web, the arrow shows the direction of energy flow.

Interesting Facts About Decomposers

In nature, when plants die, the nutrients stored in their cells are decomposed and returned to the soil. Bacteria convert nitrogen from the cells into substances called *nitrates*, which can easily be absorbed by other plants. Plants need these nitrates for healthy growth. What happens when we do not allow the plant remains to go back into the soil? When we cut our lawns, the clippings are often removed. They are not left on the ground to decompose. This does not allow the nitrogen and other nutrients to be recycled by the decomposers. In a way, by keeping our garden neat, we are interfering with the nitrogen cycle. In order to maintain a beautiful lawn and garden, we add fertilizer to the soil. The fertilizer is a chemical mix that feeds nitrogen and other nutrients back into the soil. A fertilizer has three numbers associated with it, for example 5-10-5. The first number indicates the percentage of nitrogen in the fertilizer. The other numbers indicate the percentages of two other important nutrients: phosphates and potassium.

Activity

Read the following passage and use the information provided in the table and figure to complete this activity.

A student studied the animals in an environment and reported the following: There are trees and grasses. Rabbits and deer eat the grass. Deer, squirrels, and small birds eat the fruits, nuts, berries, and leaves of the trees. Foxes hunt for the rabbits and squirrels. Large birds, such as hawks, also eat the rabbits.

1. Copy the diagram below into your notebook. Fill in each of the organisms from the food web described above. Draw the arrow heads on each line to show the direction of energy flow.

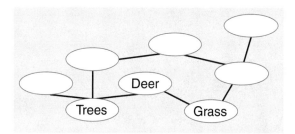

2. Copy the table below into your notebook. Based on the food web described above, identify each of the organisms as a producer, herbivore, or carnivore by checking the appropriate box. The first organism has been done for you.

Organism	Producer	Herbivore	Carnivore
Grasses	✓		
Foxes			
Rabbits			
Deer			
Small birds			
Trees			
Large birds			
Squirrels			

Questions

Refer to the following diagram, which shows a food web.

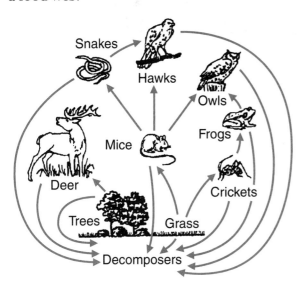

1. What can you conclude about the animals in the food web?
   (1) They all eat other animals.
   (2) They obtain their energy by eating other organisms.
   (3) They require only sunlight to survive.
   (4) They manufacture their own food

2. Which animal in this food web has the greatest number of predators?
   (1) mouse   (3) snake
   (2) frog   (4) cricket

3. Which animal's survival is **least** affected by the other animals in this food web?
   (1) mouse   (3) hawk
   (2) frog   (4) deer

4. The herbivores in this food web include
   (1) deer and mice
   (2) hawks and crickets
   (3) snakes and frogs
   (4) trees and grasses

Thinking and Analyzing

Use the food web illustrated above to answer questions 1–3.

1. List three organisms in the food web that are competing for the same food resource.

2. How many organisms in this food web feed on the grass?

3. Are deer considered herbivores, carnivores, or omnivores? Explain.

7.6 How Do Changes in a Food Web Affect the Organisms in It?

Objectives

Explain how competition affects the organisms in a habitat.

Understand how the loss of one species may affect other species in a food web.

Terms

habitat: the particular environment in which organisms live, such as a desert, a pond, a forest, or a jungle

competition: the interaction between organisms that require the same food and resources

A Delicate Balance

As we saw with the food chain, energy flows through the food web from organism to organism. An ecosystem must contain a constant supply of energy that is available to all organisms within it. Even the lives of animals that do not prey on one another may be connected. This results in a delicate balance among all the organisms in a food web. Any small change to any one of the organisms living in a habitat can affect *all* the organisms living there. A **habitat** is the particular environment where they all live together.

An example of this delicate balance can be seen in our food web in Figure 7.5-1 on page 276. Both the fox and the hawk prey on small birds and rabbits. The fox and the hawk are in **competition** for the same food items. A decrease in the number of foxes increases the amount of food available to the

hawks. Similarly, if the number of foxes increases, there is less food available for the hawks. Either of these changes affects the number of small birds and rabbits.

Upsetting the Balance

Suppose that one species is removed from an ecosystem, perhaps due to people's actions. What effect does this have on the other organisms in that system? If the rabbits are removed, for example, there is more food available to birds that compete with the rabbits. The birds benefit from the additional food. On the other hand, with no rabbits as prey, the hawks would probably eat more small birds.

What happens if the trees in which the small birds build their nests are destroyed? The birds move away. With no birds to hunt them, the number of grasshoppers increases. Eventually, the grasshoppers might eat so much of the grasses and shrubs that there is not enough food remaining to keep the rabbits alive. With no rabbits, and no small birds, the foxes and hawks might also have too little food to survive. Any small change to just one member of the ecosystem may have a tremendous effect on all of them.

Interesting Facts About Lions

To avoid their predators, prey animals are able to run very fast. Many of the animals that lions eat, such as deer, antelopes, and gazelles, can run 40, 50, or even 60 miles per hour. A lion, however, can run only 35 miles per hour, much slower than its prey (but much faster than you or I!). A lion must sneak up on its prey to catch it. It usually catches the slowest animals in the herd, the sick or elderly. When the prey is about 50 feet away, the lion rushes forward and grabs the animal, usually biting its neck. The lion can open its mouth 11 inches and kill an antelope or zebra in one bite. Lions eat every 3 to 4 days, and can eat 75 pounds of meat at one meal. That would be 300 quarter-pounders—without the buns.

Questions

Use the illustration of the food web at right to answer questions 1 and 2.

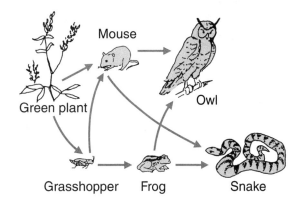

1. Which animal is in competition with the snake?
   (1) frog   (3) owl
   (2) mouse   (4) grasshopper
2. An increase in the number of mice might first result in a decrease in
   (1) frogs   (3) owls
   (2) grasshoppers (4) snakes

Thinking and Analyzing

Refer to the food web that you constructed for the Activity in Lesson 7.5 on page 278 to answer questions 1 and 2.

1. Explain what happens to the number of squirrels if a disease destroys the fruits and nuts in the ecosystem.
2. What happens to the other organisms in the food web if a harsh winter kills all the rabbits?

Refer to the food web above to answer questions 3 and 4.

3. How would the grasshopper population be affected if the frogs disappear? Explain your answer.
4. How would the mice population be affected if the frogs disappear? Explain your answer.

What Happens to the Energy in a Food Web?

Objectives

Create an energy pyramid from a food chain or food web.

Compare the energy available at each layer in an energy pyramid.

Identify primary consumers and secondary consumers.

Terms

energy pyramid: a device used to show the energy relationships between producers and consumers in a food chain; the producers are always at the base of an energy pyramid

primary consumer: an animal that eats producers; a herbivore

secondary consumer: an animal that eats herbivores

Loss of Energy

What happens to you when you exercise? When you exercise, your body burns food for energy. As you exercise, your body temperature increases. Heat is being created and lost to the environment. Every organism uses some of the energy it consumes and stores the rest. The energy that is used is lost to the environment in the form of heat. Once energy is used, it cannot be replaced. Only the stored energy is available to the next consumer. This results in less and less energy available as energy moves from organism to organism in a food web. You may recall the law of conservation of energy (see Lesson 2-7), which states that energy can neither be created nor destroyed. What do we mean when we say that energy is "lost"? We are referring to energy that is no longer available to organisms in the food chain. ***Stop and Think:*** Which of the organisms in Figure 7.5-1 on page 276 store the greatest total amount of energy? ***Answer:*** The producers always contain the greatest amount of energy in any food web.

Primary Consumers

An **energy pyramid** shows the loss of energy through a food web. Look at the energy pyramid in Figure 7.7-1. Because the producer layer contains the most energy, it is always the widest part of the pyramid. The grasshopper gets its energy directly from the producers. Animals that get their energy directly from producers are called **primary consumers.** ("Primary" means "first.") All herbivores are primary consumers.

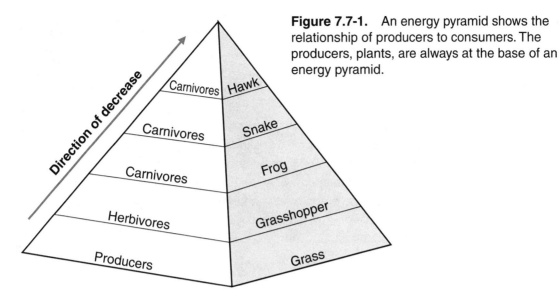

Figure 7.7-1. An energy pyramid shows the relationship of producers to consumers. The producers, plants, are always at the base of an energy pyramid.

The grasshopper and the other primary consumers get their energy from green plants. They use most of this energy in carrying out their life processes, but they store some energy as well. The amount of energy stored by the herbivore is always less than the amount of energy that was in the food they consumed. Therefore, primary consumers store less energy than producers.

Secondary Consumers

The carnivores that eat the herbivores—the primary consumers— are called **secondary consumers**. The frog in Figure 7.7-1 is a secondary consumer. The grass, the grasshopper, and now the frog each use up some of the energy originally provided by the sun. Only a small portion of this energy is now available to the rest of the food chain. The amount of available energy decreases with each additional layer in the energy pyramid. Scientists estimate that each layer contains ten percent of the energy in the level below it.

The loss of energy as we move up the energy pyramid limits the number of consumers in each layer. Therefore, there are always more producers in an ecosystem than there are primary consumers. There are more primary consumers than secondary consumers. If the number of organisms at any level of the pyramid grows too large, there will not be enough food available for all of them.

Activity

Draw and label an energy pyramid based on the organisms shown in Figure 7.5-2 on page 277.

Interesting Facts About Energy Pyramids

You may have noticed that in our energy pyramids, the largest animals are at the top. This is not always the case in nature. As a matter of fact, the largest animals are usually herbivores. In Africa, for example, lions are the largest carnivores, but they are much smaller than elephants, giraffes, gnus, and hippopotami, which are all herbivores. Large animals require a great deal of energy to sustain them. It takes more energy to move a heavy object than it takes to move a light object. Moving an elephant requires more energy than moving a lion. Remember, the largest amount of available energy is at the base of the pyramid, available only to herbivores. The predators are near the top of the pyramid, where there is less available energy. If a predator is too large, it will not be able to find enough food to meet its energy needs.

(Try to imagine the amount of energy consumed by a herd of stampeding elephants. That sure requires a lot of peanuts!)

Questions

Base your answers to questions 1–4 on the energy pyramid at right.

1. The greatest amount of energy is stored by the
   (1) sharks
   (2) phytoplankton
   (3) mackerel
   (4) herring

2. The primary consumers in this pyramid are the
   (1) sharks
   (2) small crustaceans
   (3) mackerel
   (4) phytoplankton

3. The organism in this ecosystem that is present in the **smallest** numbers is probably the
   (1) shark
   (2) mackerel
   (3) herring
   (4) phytoplankton

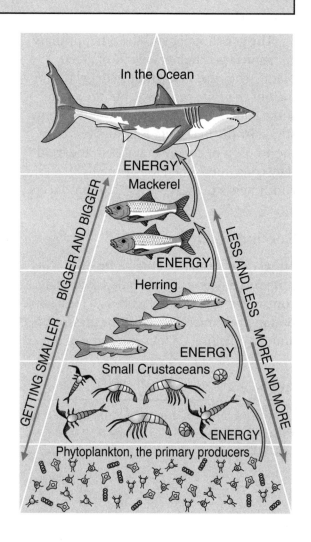

In the Ocean

BIGGER AND BIGGER

GETTING SMALLER

LESS AND LESS

MORE AND MORE

ENERGY
Mackerel
ENERGY
Herring
ENERGY
Small Crustaceans
ENERGY
Phytoplankton, the primary producers

4. Which organism is considered a producer?
   (1) the shark, because it is at the top of the pyramid
   (2) the small crustaceans, because they change the energy in plants into food
   (3) the phytoplankton, because they change energy from the sun into food
   (4) the mackerel and the herring, because they both can be eaten

5. In the food web shown in Figure 7-5.1 on page 276, the **smallest** amount of energy is available from the
   (1) grasses and shrubs
   (3) rabbits and grasshoppers
   (2) small birds
   (4) foxes and hawks

Thinking and Analyzing

1. Some birds are primary consumers and some are secondary consumers. You can sometimes tell which they are just by looking at the shapes of their beaks. Visit the website *http://mdc.mo.gov/kids /out-in/ 2006/02/1.htm* to learn about the feeding habits of different birds. Give one example of a bird that is a primary consumer and one that is a secondary consumer. Explain your answer.

2. Where is the greatest amount of energy stored in an energy pyramid?

3. Explain why a human being is considered both a primary and a secondary consumer.

4. Cows eat grass, and humans eat cows. Draw an energy pyramid that represents this food chain. Where in this pyramid is the greatest amount of energy? Where is the least?

7.8 How Do Green Plants Make Energy Available to Consumers?

Objectives

Describe the process of photosynthesis.

Identify the materials needed for photosynthesis.

Identify the materials produced from photosynthesis.

Explain how plants store food.

Terms

photosynthesis (foe-toe-SIN-thuh-sis): the chemical process in which a green plant uses sunlight to convert carbon dioxide and water into glucose and oxygen

glucose (GLEW-kose): a simple sugar produced during photosynthesis

chlorophyll (KLOR-uh-fill): a green chemical in plants needed for photosynthesis

starch: a nutrient produced from sugar that stores energy in plants

Photosynthesis

As we have already learned, animals obtain nutrients by consuming other living things. Green plants, on the other hand, manufacture their own food through a process called **photosynthesis**.

During the process of photosynthesis, plants use the energy from sunlight to change carbon dioxide and water into **glucose**, a type of sugar. (See Figure 7.8-1.) The leaves of the plant absorb carbon dioxide from the air, while the roots of the plant absorb water from the soil.

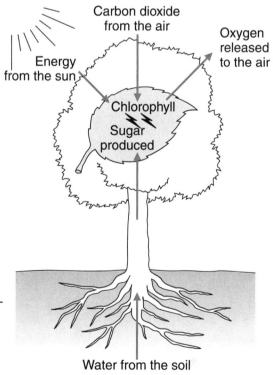

Figure 7.8-1. The basic process of photosynthesis—plants take in water and carbon dioxide from the environment and produce glucose (sugar) and oxygen.

Photosynthesis takes place in the leaves of trees and other green plants. The green color indicates the presence of **chlorophyll**, a chemical involved in photosynthesis. Photosynthesis requires all four components: chlorophyll, water, carbon dioxide, and sunlight.

Photosynthesis may be summarized using the following chemical equation:

$$6CO_2 + 6H_2O + energy \rightarrow C_6H_{12}O_6 + 6O_2$$
$$carbon\ dioxide + water + energy \rightarrow$$
$$glucose + oxygen$$

In addition to glucose, photosynthesis also produces oxygen, which plants release into the environment. The process of photosynthesis in green plants produces much of the oxygen we breathe on Earth.

Storing Energy

The glucose produced through photosynthesis is needed by the plant for energy to carry out life processes. Some of this sugar must be stored for use at a later time. Different plants store sugar in different ways. The potato plant, for example, converts sugar into **starch**, which it stores in a potato. The maple tree, on the other hand, stores sugar in its sap, which we use to make maple syrup.

Obtaining Energy from Plants

Animals that eat plants use the stored sugar to produce energy. However, animals cannot obtain energy directly from starches and other stored sugars. First, the process of digestion changes these substances back to glucose. Glucose and oxygen are transported to each and every cell where they combine to produce energy during respiration. *Respiration* can be summarized using the following chemical equation:

$$C_6H_{12}O_6 + 6O_2 \rightarrow 6CO_2 + 6H_2O + energy$$
$$glucose + oxygen \rightarrow carbon\ dioxide$$
$$+ water + energy$$

Compare this equation to the one for photosynthesis. They are opposites!

Activity

The diagram illustrates a leaf of a plant that has been covered with a piece of black paper for several days. When the black paper is removed, the area covered by the paper no longer looks green. When this leaf is tested for starch, the uncovered portion tests positive (starch is present), and the covered portion tests negative (no starch is present.) Explain why the covered and uncovered portions of the leaf have different test results.

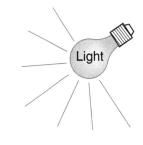

Light

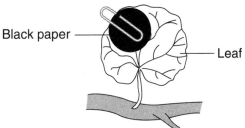

Black paper

Leaf

Interesting Facts About Phytoplankton

Phytoplankton can be seen from outer space

Phytoplankton are microscopic plant cells that are the basis of all the ocean's food chains. These plants grow in large clusters near the surface of the ocean. Although these plants are so small that the individual organisms cannot be seen without a microscope, there are so many of them that they can be seen from space.

Scientists use satellite images to keep track of concentrations of phytoplankton in our oceans. The satellites can detect the green color from the phytoplankton's chlorophyll. Scientists noticed a decline in the amount of phytoplankton in the North Atlantic Ocean in the 1980s and 1990s. Scientists are concerned because these organisms are the source of energy for most of the ocean's food chains. Phytoplankton also produce oxygen and remove carbon dioxide from the atmosphere.

The plants rely on sunlight and on other nutrients that come from deeper levels in the ocean. Scientists suspect that warmer ocean temperatures and calmer winds prevent the oceans layers from mixing, keeping nutrients at lower levels and out of reach of the phytoplankton.

But don't worry! More recent measurements show that the trend may be reversing, and the number of phytoplankton is beginning to increase.

Questions

1. Which observation indicates that photosynthesis is taking place in a plant?
   (1) The flowers have a sweet odor.
   (2) The stem is made of wood.
   (3) The leaves are green.
   (4) The roots grow deep into the ground.

2. Which of the following is **not** needed for photosynthesis?
   (1) water
   (2) oxygen
   (3) carbon dioxide
   (4) sunlight

3. The products of photosynthesis are sugar and
   (1) carbon dioxide
   (2) water
   (3) salt
   (4) oxygen

4. If a plant lost its ability to make starch, it would no longer be able to
   (1) take in carbon dioxide
   (2) release oxygen gas
   (3) store extra sugars
   (4) undergo photosynthesis

5. The process that a producer uses to make its own food is called
   (1) photosynthesis
   (2) digestion
   (3) eating
   (4) respiration

6. The chemical reaction that obtains energy from glucose occurs during the process of
   (1) photosynthesis
   (2) digestion
   (3) respiration
   (4) excretion

Thinking and Analyzing

Base your answers to questions 1 and 2 on the paragraph below.

Joseph's aquarium contains water, gravel, fish, green plants, and a source of light. He adds fish food daily. He observes tiny bubbles forming on the leaves of the plants. He wonders how these bubbles were formed and what happens to them. He also notices that when he leaves the light on for a longer period of time, the number of bubbles increases.

1. What gas is responsible for the bubbles formed on the leaves, and how is this gas formed?

2. What might happen to the plants if the light were turned off for a long period of time?

3. Describe the relationship between photosynthesis and respiration.

4. During the winter, maple trees have no leaves, and so no photosynthesis can occur. Where do the trees obtain the energy they need to produce leaves again the following spring?

Review Questions

Term Identification

Each question below shows two terms from Chapter 7. One of the terms is defined.
(1) Choose the term that matches the definition.
(2) Describe how the two terms are different.
Following each term is the section (in parenthesis) where the description or definition of that term is found.

1. *Ecology (7.1) — Environment (7.1)*
   The branch of biology that studies the relationships between living things and their surroundings

2. *Omnivore (7.3) — Carnivore (7.3)*
   An animal that obtains its energy only by eating other animals

3. *Predator (7.3) — Prey (7.3)*
   An animal that is hunted by other animals

4. *Producers (7.4) — Consumers (7.4)*
   Organisms that obtain their energy from other organisms

5. *Primary consumer (7.7) — Secondary consumer (7.7)*
   A herbivore

6. *Glucose (7.8) — Chlorophyll (7.8)*
   A sugar that is produced in photosynthesis

Multiple Choice (Part 1)

Choose the response that best completes the sentence or answers the question.
Use the illustration below to answer questions 1 and 2.

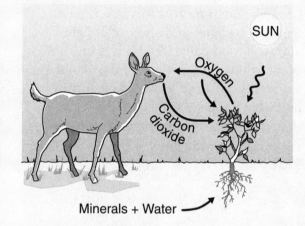

Minerals + Water

1. Which statement best explains the relationships shown?
   (1) Over time, water is transferred from one organism to another.
   (2) Living things exchange materials with their environment.
   (3) Minerals recycle the dead materials in the environment.
   (4) Living things produce other living things.

2. Which process produces oxygen that is released into the atmosphere?
   (1) respiration   (3) excretion
   (2) locomotion   (4) photosynthesis

Use the illustration below to answer questions 3 and 4.

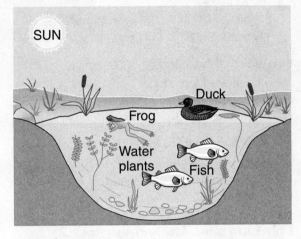

3. Why are the fish able to survive in the pond?
   (1) The fish use carbon dioxide produced by the plants.
   (2) The fish use oxygen produced by the plants.
   (3) The plants use oxygen produced by the fish.
   (4) The plants use chlorophyll produced by the fish.

4. The main source of energy for this pond community is the
   (1) plants   (3) pond water
   (2) sun   (4) animals

Base your answers to questions 5–10 on the figure below and your knowledge of food webs.

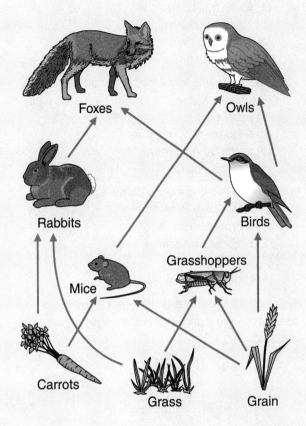

5. What is the role of the foxes in this food web?
   (1) carnivore   (3) herbivore
   (2) decomposer   (4) producer

6. Which organisms in this food web are omnivores?
   (1) birds   (3) grasshoppers
   (2) carrots   (4) rabbits

7. Which statement is true if the owl population disappears?
   (1) The mouse population increases.
   (2) The carrot population increases.
   (3) The fox population decreases.
   (4) Other animals in the food web would not be affected.

8. Which organisms compete for the same food source?
   (1) foxes and mice
   (2) rabbits and birds
   (3) foxes and owls
   (4) grasshoppers and owls

9. This food web leaves out a group of organisms known as
   (1) producers
   (2) secondary consumers
   (3) primary consumers
   (4) decomposers

10. Which two organisms demonstrate a predator-prey relationship?
    (1) carrots and mice
    (2) birds and grasshoppers
    (3) foxes and owls
    (4) foxes and mice

Base your answers to questions 11 and 12 on the figure below and your knowledge of energy pyramids.

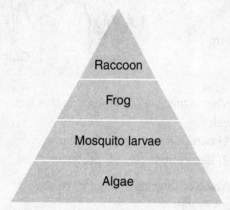

11. Which term best describes the mosquito larvae in this energy pyramid?
    (1) producer   (3) carnivore
    (2) predator   (4) consumer

12. Which of the following organisms does **not** provide energy for any of the other organisms in this energy pyramid?
    (1) algae   (3) frogs
    (2) mosquito larvae (4) raccoons

13. Which of the choices below includes the other three?
    (1) air   (3) environment
    (2) water   (4) organisms

14. Which of the following organisms is considered a producer?
    (1) bumble bee   (3) apple tree
    (2) hummingbird   (4) kangaroo

15. Which of the following is **not** required by a plant to make sugar?
    (1) carbon dioxide   (3) water
    (2) chlorophyll   (4) oxygen

Thinking and Analyzing (Part 2)

Base your answers to questions 1–5 on the figure below and your knowledge of food webs.

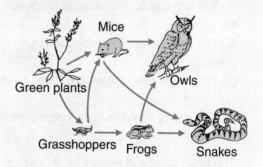

1. Explain how the feeding habits of mice are different from the feeding habits of the other organisms shown in this food web.

2. Explain this statement: "The energy for all the organisms in this food web can be traced back to the sun."

3. Give an example of a carnivore, a producer, and an herbivore shown in this food web.

4. Identify a food chain in this food web that consists of four organisms.

5. Draw an energy pyramid using the food chain you identified in question 4.

Base your answers to questions 6–9 on the figure at right and your knowledge of ecosystems.

6. What is the main source of energy for this ecosystem?

7. What is the name of the process that converts this energy into food?

8. Identify a consumer in this ecosystem.

9. Identify **one** organism in this ecosystem that produces chlorophyll.

Chapter Puzzle (*Hint:* The words in this puzzle are terms used in the chapter.)

Across

1 an organism that eats other organisms

5 a device used to show the energy relationships in a food chain is called an energy _____

7 an organism that makes its own food

10 the type of organism that carries out photosynthesis

12 the source of all the energy in a food chain

13 the study of the interaction between organisms and their environment

15 a plant-eating animal

16 a meat-eating animal

17 an animal that eats the remains of dead animals

18 what the predator hunts

Down

1 the flow of nutrients or energy from one organism to another is called a food _____

2 one of the products of photosynthesis (also called glucose)

3 the surroundings in which an organism lives

4 an animal that hunts and kills for food

6 a system formed by the interaction of organisms with their physical environment

8 a special organism, such as a fungus, that breaks down the remains of dead organisms

9 a nutrient produced from sugar that is used by plants to store energy

11 several connected food chains are called a food _____

14 an animal that eats both plants and animals

16 a series of events that is repeated over and over

Unit 4

Interdependence

Part I—Essential Question

This unit focuses on the following essential question:

How is interdependence essential in maintaining life on Earth?

In Unit 4, we focus on ecology, the study of the interaction between organisms and their environment. One factor that sets one environment apart from another is climate. In our study of climates, we consider many of the factors we discussed in Unit 2, the weather unit. We will illustrate the patterns of temperature, wind, and precipitation in many different parts of the world. You will learn about New York's climate, and many others climates, including the dry deserts and wet rainforests.

Different climates support different types of living things. Places that have similar climates, soil, and plant life, generally support similar animal life as well. A particular combination of climate, soil, and plant life is called a biome. We will study the interdependence of these three factors within several biomes.

The living and non-living parts of an environment form an ecosystem. Within each ecosystem, organisms are dependent for their survival on other organisms. This unit examines the interdependence of the plants and animals in an ecosystem.

Part II—Chapter Overview

Chapter 8 is called "Climate and Biomes." This chapter focuses first on the different climates of the world – how they are different, and why they are different. We will show how latitude, elevation, ocean currents, and wind currents combine to determine a climate. Once we have described several of Earth's climates, we will examine the different biomes that are found in some of those climates.

Climates can change over time. Earth's climate, according to most scientists, has been getting warmer for the past 200 years. This change, called global warming, seems to becoming more rapid. Global warming, what is causing it, how it may affect us in the future, and whether we can do anything to prevent it, are all topics being debated by scientists around the world. This important issue is discussed in Lesson 8.4.

Chapter 9 is called "Ecosystems and Interdependence." In this chapter, you will become familiar with the vocabulary of ecology, as you learn how ecologists use terms like "population" and "community."

You will see how some of the populations in your community, and in other communities, interact with each other.

Chapter 9 examines some of the special relationships between interdependent organisms. Some organisms depend on you for their survival, but you probably are not interested in helping them! These include parasites, such as fleas, ticks, and bacteria. As you will see, there are other relationships between different organisms that are beneficial to both of them.

In Chapter 9, we discuss how environments may change over time. As we mentioned above, one factor, global warming, is covered in Chapter 8. However, we will see that there are several other causes of environmental change. These include natural events, such as volcanic eruptions, floods or droughts, and many human activities, such as building dams and clearing forests. We will study how some of these changes affect us, and the other living things around us.

Chapter 8

Climates and Biomes

Contents

The African savannah is home to many plants and animals.

What Is This Chapter About?

Our planet is made up of many different types of environments that change over time. These changes are a result of natural as well as human activities.

In this chapter you will learn:

1. What climate is, and what factors determine the climate for a location.
2. How climate affects the species of organisms in a particular location.
3. How environments are classified.
4. How environments change over time.
5. How and why Earth's climates may change in the future.

Career Planning: Climatologist

Who is keeping track of the climate? The answer is, everyone! In the mid 1960s, scientists suggested that human activity might be causing changes in our climate that could affect the lives of many people in the future. Scientists started studying one of these changes, called global warming, very carefully. Climatologists are one group of scientists who study global warming. The work of a climatologist involves studying Earth's climatic changes. Using sediments from lakes, ice taken from deep in the polar ice caps, ocean temperatures, and even growth rings in tree trunks, climatologists learn what changes have occurred to Earth's climate. With this information and the help of computer modeling, they try to predict future changes to the climate.

Internet Sites:

http://www.geography4kids.com/files/climate_intro.html Visit the climate section of this great Web site. Make sure you watch all the videos at the bottom of the page.

http://www.mbgnet.net Visit the biomes of the world and see the plants and animals that live in them.

http://www.epa.gov/globalwarming/kids/greenhouse.html At this Web site for kids by the Environmental Protection Agency (EPA), learn about global warming and the greenhouse effect.

http://tiki.oneworld.net/penguin/global_warming/climate_home.html Climate and climate change are illustrated and explained at this Web site.

8.1 What Factors Cause Differences in Climate?

Objectives

Identify temperature and precipitation as two factors in determining climate.

Identify three climate zones based on latitude.

Understand how latitude and elevation affect climate.

Identify factors that affect climate other than latitude and elevation.

Terms

climate: the average weather in an area

tropical: a hot and moist climate

temperate (TEHM-puhr-iht): a climate with warm summers and cold winters

equator (ee-KWAY-tuhr): an imaginary line around the center of Earth midway between the North and South Poles

latitude (LAT-ih-tude): a measure of distance north or south of the Equator, expressed in degrees

elevation (eh-luh-VAY-shun): the height above sea level

What Is Climate?

Have you ever heard the rumor that there are alligators living in the New York City sewer system? Do you believe it? Alligators are common in Florida, but it is very unlikely that one could survive in New York City. There are many organisms that are able to survive in one part of the country, but not in another. Alligators can survive in Florida but not in New York because Florida and New York have different climates.

Climate is the average of all weather conditions of a particular area. It tells us how much precipitation and what temperatures we might expect during the course of a year.

Climate is determined by studying the weather over a long period of time. Florida has a **tropical** climate. This means that the temperature remains warm all year. New York has a temperate climate. A **temperate** climate has warm summers and cold winters.

Latitude

What accounts for the difference in climate between Florida and New York? You probably know that in the United States, weather generally gets warmer as you go south. The sun does not heat Earth evenly. Heating is strongest near the equator, and weakest at the poles. The **equator** is an imaginary line around Earth midway between the North Pole and the South Pole. (See Figure 8.1-1.) In North America, as you go south, you are getting closer to the equator.

Near the equator, there are only small, seasonal changes in the weather. The farther you travel north or south of the equator, the greater the differences between summer and winter.

Latitude is a measure of how far a place is from the equator. It is measured in degrees north (N) or south (S). The equator is given a latitude of 0°. The North Pole is located at 90°N latitude and the South Pole at 90°S latitude. New York City is at 40°N, while South Florida is at 25°N. Figure 8.1-2 shows how degrees of latitude are determined.

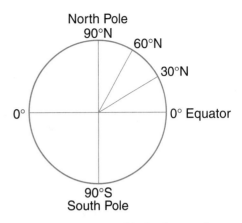

Figure 8.1-2. Degrees of latitude are the angles measured from the center of Earth and the equator to another point on Earth.

Climate Groups

Scientists have identified three climate groups based on latitude. (See Figure 8.1-1.) Low-latitude climates are located nearest the equator. These climates remain warm all year. In the United States, only Hawaii and the southern tip of Florida fall into this group. Most of the United States is in the mid-latitude group. These climates usually have a warm summer and a cold winter. Alaska is the only American state in the high-latitude group. These climates typically have extremely cold winters. Near the North and South poles, it is so cold that even the oceans are frozen solid.

Factors That Determine Climate

As you can see in Figure 8.1-3 on the next page, the latitude of San Francisco, California, is about 3° south of the latitude of New York City. As a result, you might expect it to be warmer there. Although San Francisco is much warmer than New York in the winter, New York is

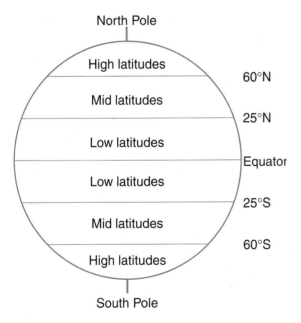

Figure 8.1-1. Earth is separated into three climatic groups, starting at the equator and moving north and south toward the poles.

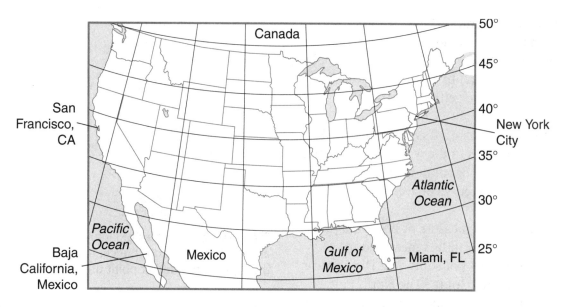

Figure 8.1-3. Miami in Southern Florida and Baja, California are at the same latitude. New York City and San Francisco are almost at the same latitude.

warmer than San Francisco in the summer. Latitude is not the only factor that determines climate. Global winds and ocean currents also affect the climates of New York City and San Francisco. Table 8.1-1 lists some of the factors that determine climate.

Types of Climates

Southern Florida has a tropical climate. Baja California, in Mexico, is at the same latitude (see Figure 8.1-3), but has a very different type of climate. The difference is due to the amount of rainfall each receives. Florida has

Interesting Facts About Climate

Are you interested in gardening? If you want to order plants or seeds, you need to know what plants will grow in your climate. To help you, plant catalogs list a hardiness zone with the description of each plant. Most of New York City is in hardiness zone 6, while Miami is in hardiness zone 10. The lower the number is, the colder the climate. Would you like to grow gardenias? A gardenia is a very fragrant white flower. Unfortunately, it grows only in hardiness zones 7–10, that is, in warmer climates. You can grow lilacs, though. They grow in hardiness zones 3 to 8. Gardening catalogs list thousands of different plants. If you know your hardiness zone you can find out not only what to plant, but what time of year to plant it. To see a map of hardiness zones for the eastern United States, visit *http://www.boldweb.com/greenweb/g_zoneus.gif*

Table 8.1-1. Factors That Determine Climate

Factor	How It Affects Climate	Example
Latitude	High latitudes have cooler climates than low latitudes.	The coldest temperatures in the U.S. have been recorded in Alaska, the northernmost state.
Elevation	High elevations have cooler climates than low elevations.	At 14,410 feet above sea level, Mount Rainier is covered with snow all year round.
Large Bodies of Water	Land areas close to large bodies of water experience a smaller variation between seasonal high and low temperatures.	New York City is located near the Atlantic Ocean and is cooler in the summer than Pittsburgh, which is inland.
Mountain Barriers	A mountain can cause rain to fall on one side, keeping its other side dry.	Death Valley, located east of the Sierra Nevada Mountains, is one of the driest places in the U.S.
Ocean Currents	Warm ocean currents carry heat to cooler regions. Cool ocean currents cool warmer regions.	A cold Pacific Ocean current makes summers on the west coast of the U.S. cooler than summers on the east coast.
Global Winds	Global winds move warm and cold air over great distances.	The prevailing winds over the U.S. blow west to east.

a wet climate, while Baja California has a dry climate. (See Figure 8.1-4 on the next page.) Climates are classified on the basis of *both* temperature and precipitation.

Recall that New York City has a temperate climate; it has warm summers and cold winters. Just as there are different types of tropical climates, there are different types of temperate climates. New York City's climate is a moist, continental climate. It is called moist, because it receives an average of 120 centimeters (47 inches) of rainfall per year. It is called continental, because its climate is controlled mainly by air masses that pass over land. On average, areas within this climate type have between 75 and 150 centimeters of rainfall each year. They have an average yearly temperature of about 50°F (10°C).

The west coast of the United States has a Mediterranean climate. It has moist winters and dry summers. San Francisco in California receives an average of 45 cm (17.6 in.) of rain from November through April each year, but averages only 5 cm (2 in.) of rain for the rest of the year.

Figure 8.1-4. (Left) Cactus in Baja, California—a desert climate. (Right) Cypress trees in Florida—a wet, tropical climate.

Elevation

Look at Figure 8.1-5, a photograph of Mount Kilimanjaro in Africa. Mount Kilimanjaro is a snow-capped peak located very close to the equator. You might be surprised to find snow there all year long. Mount Kilimanjaro is the highest peak in Africa, at over 19,000 feet above sea level. At such a high elevation, it is common to find snow-covered mountains, even at the equator.

Elevation refers to how high above sea level a place is. As you climb higher from ground level, the average temperature decreases. This results in cooler climates at higher elevations. If you climb high enough, you may enter a completely different type of climate.

Figure 8.1-5. Snow-covered peak of Mount Kilimanjaro in Africa.

Activity

What type of climate do you like to visit on vacation? Some people like to swim on vacation, while others prefer to ski down a mountain. Use the Web site *http://www.worldclimate.com* and look up a city you would like to visit on vacation and describe its climate.

1. What city would you like to visit?
2. What is the average annual maximum temperature?
3. Does the temperature change much from summer to winter?
4. How much rainfall does it get in a year?
5. Are there wet and dry seasons?
6. What is its latitude?
7. What is its elevation?

Questions

1. New York City has a cooler climate than Miami, Florida. This difference is mainly due to differences in
   (1) distance from the ocean
   (2) elevation
   (3) latitude
   (4) mountain barriers

2. In which climate group is New York City found?
   (1) low latitude   (3) high latitude
   (2) mid latitude   (4) bad latitude

3. While it often rains in Seattle, Washington, it rarely snows. However, Mount Rainier, located just outside of Seattle, is always covered with snow. The difference in climate between Seattle and Mount Rainier is due to differences in
   (1) distance from the ocean
   (2) elevation
   (3) latitude
   (4) ocean currents

4. Climate is based on
   (1) only the average annual temperature
   (2) only the average annual rainfall
   (3) neither the average annual temperature nor the annual average rainfall
   (4) both the average annual temperature and the average annual rainfall

Thinking and Analyzing

1. Why is the climate in San Francisco, California different from the climate in Baja California, Mexico?

2. Why is the climate in New York different from the climate in California?

3. What do you expect to happen to the temperature as you climb a mountain?

How Are the World's Environments Organized?

Objectives

Identify several factors that determine a biome.

Compare and contrast major biomes.

Terms

desert (DEH-zurt): a biome that receives less than 25 centimeters (10 inches) of rain per year

biome (BUY-ohm): a region of Earth that is characterized by a particular climate, soil, and community of plants

deciduous (dee-SID-you-us): a type of tree or shrub that drops its leaves in the winter

temperate deciduous forest: a biome containing deciduous trees

temperate grassland: a biome that contains tall grass and few trees

savannah (suh-VAH-nuh): a tropical grassland

tropical rain forest: a biome with a warm, wet climate in which the climax community contains broadleaf evergreens

tundra (TUHN-druh): a treeless biome with soil that is always frozen

taiga (TAY-guh): a biome with long, cold winters that contains needle-leaf evergreen trees, such as fir and spruce

Deserts

Where do you expect to find a cactus growing? As you probably already know, a cactus is a plant that survives in an area that may have high temperatures and very little water. An area like this is called a **desert**. Other plants, such as oak trees and pine trees, do not survive in this type of habitat because they require a large amount of water. A desert habitat receives less than 25 centimeters (10 inches) of rain per year. Figure 8.2-1 shows some desert plants and animals.

Figure 8.2-1. Desert plants and animals.

Biomes

A desert is just one of many types of environments found on Earth. These environments may differ in climate, temperature, soil, moisture, and the amount of sunlight they receive. These are some of the factors that determine which communities of plants and animals are found in a given environment. Scientists organize Earth into several major groupings that describe the environments and the communities that live in them. These groupings are called **biomes**. A biome has many *ecosystems* within it.

A desert is a land biome. Other land biomes include grasslands and tropical rain forests. (See Table 8.2-1 on page 309 for a list and description of several of the world's biomes.) Notice that a particular biome is usually located at a particular *latitude* on Earth (both north and south of the equator). This is because climate is one of the major factors that determines a biome.

Temperate Deciduous Forest

You probably do not live in a desert or tropical rain forest. The biome you live in probably has four distinct seasons: summer, winter, spring, and fall. Some animals in your biome have fur to keep them warm in winter, while others may migrate to warmer climates. There are tall trees whose leaves change color in the autumn and fall off in the winter. Trees that lose their leaves in the winter are called **deciduous** trees. They live in a biome called a **temperate deciduous forest**. Most of the eastern half of the United States is a temperate deciduous forest. Figure 8.2-2 shows the types of

Figure 8.2-2. Plants and animals of the temperate deciduous forest.

trees and animals that live in temperate deciduous forests.

Grasslands

The central portion of the United States looks very different from the forests of the east. Much of the central United States is flat without many trees. The land is covered with grass. There are dry seasons and wet seasons, but not enough water to grow big trees. This biome is called a **temperate grassland**. Most of these areas have been cleared for farming and no longer support the large numbers of animals that once lived there. Much of Africa is a biome called a **savannah**. A savannah is a *tropical grassland*. The grasslands of Africa still support large herds of grazing animals and the predators that feed on them. (See Figure 8.2-3 on the next page.)

Figure 8.2-3. Plants and animals of the savannah.

Tropical Rain Forest

Another biome you may have heard of is the **tropical rain forest**. Broadleaf evergreen trees characterize this biome, as shown in Figure 8.2-4. These trees stay green all year. It is usually warm all year, and there is plenty of water available. The tropical rain forests are home to many species of plants and animals that are found nowhere else on Earth.

Figure 8.2-4. Plants of the tropical rain forest.

Interesting Facts About Deserts

When you think of a desert, you probably picture a very hot place with lots of sand and very little water. The habitat that you are picturing is a desert, but there are many other types of deserts. Not all deserts are covered with sand. Many are mountainous or rocky. As a matter of fact, not all deserts are hot. The one thing they do have in common, however, is that they are all dry. A desert receives less than 25 cm (10 inches) of rainfall a year. In a hot and dry desert, the rate of evaporation exceeds the rate of precipitation (rainfall).

Hot and dry deserts are located in two bands across Earth near the Tropic of Cancer and the Tropic of Capricorn. This includes a great desert in the southwestern United States. The desert in the United States is the fifth largest desert in the world, at about 500,000 square miles. The Sahara Desert in Africa is the largest hot desert on Earth, and covers almost 3.5 million square miles. The Australian deserts cover about 1.3 million square miles; the Arabian deserts cover 1 million square miles; and deserts of Turkestan cover 750,000 square miles. Hot and dry deserts cover about one-fifth of all the land on Earth.

The frozen continent of Antarctica is actually the largest desert in the world. It is about 5.5 million square miles. Interior Antarctica has only about 5 cm (2 inches) of rainfall a year. Antarctica is just one of many cold deserts. A cold desert has very long, cold winters and short summers. Other cold deserts can be found in Greenland and even in Utah.

Learn more about deserts and other biomes at these Web sites:

http://www.oswego.org/testprep/ss5/d/desert.cfm

http://curriculum.calstatela.edu/courses/builders/lessons/less /biomes/desert/cold-des.html

http://www.nceas.ucsb.edu/nceas-web/kids/biomes/biomes _home.htm

Tundra and Taiga

What are the biomes of the colder climates? Recall from Lesson 8.1 that high-latitude climates have extremely cold winters. In some of these climates, the soil is frozen all year long. The **tundra** is a biome characterized by frozen soil. Trees cannot take root in the frozen soil. Only grasses and small shrubs are able to survive. (See Figure 8.2-5 on the next page.) Not many species of animals can survive these harsh conditions.

The northern part of the United States and most of southern Canada contain a

Figure 8.2-5. Plants and animals of the tundra.

biome called the **taiga**. The taiga contains forests consisting mostly of conifers. These are trees that have needlelike leaves that stay green all year.

Table 8.2-1 lists and describes several land biomes, some of their characteristics, and their locations on Earth. The land biomes described in Table 8.2-1 describe most of the environments found on land.

Most of Earth's surface, however, is covered with water. Aquatic (watery) biomes include freshwater biomes such as lakes, ponds, and rivers. Saltwater biomes include coastal waters, open oceans, and deep oceans.

Alpine Tundra

In Lesson 8.1 you learned about the effect of elevation on climate. As you move up a mountain, temperature tends to decrease. While most of New York State contains temperate deciduous forests, the higher elevations of the Adirondack Mountains contain evergreens (conifers). Climates located at high elevations resemble those at high latitudes. As you move up a mountain, the climate may get too cold to support any trees at all. At these elevations, the plants and animals resemble those found in the tundra.

When tundra occurs due to high elevations, it is called *alpine tundra* (alpine means "of the mountains"). In the high mountain ranges, trees grow at the lower elevations, but disappear above higher elevations. The highest point where trees can grow is called the tree line.

Activity

Lions and tigers and bears, oh my! Use the Internet to determine if African lions, Bengal tigers, and grizzly bears are found in the same biome. In which biome can each of these animals be found? Some information about biomes and animals can be found at:

http://www.blueplanetbiomes.org/table_of_contents.htm
http://www.u-46.org/library/Biomes.htm
http://library.thinkquest.org/11922/habitats/habitats.htm

Table 8.2-1 Some Land Biomes

Biome	Water	Temperature	Plants	Animals	Location
Deserts	Very little	Usually hot days and cold nights	Very few; cactus	Insects, spiders, reptiles, birds, rodents that burrow during the day	Desert
Tundra	Dry	Cold	Lichens and mosses	Migrating animals	Tundra
Taiga (coniferous forest)	Adequate	Cool year-round	Conifers (needle leaf evergreens)	Many mammals, birds, insects, spiders	Taiga
Temperate Deciduous Forest	Adequate	Four seasons	Deciduous trees	Many mammals, birds, reptiles, insects, spiders	Temperate Deciduous Forest
Grasslands (both temperate and tropical)	Wet season, dry season, but not enough water for large trees	Temperate: warm and cold seasons; tropical: warm, little variation in temperature	Grasses (few or no trees)	Many mammals, birds, insects, spiders	Grasslands
Tropical Rain Forest	Very wet	Always warm	Many plants	Many animals	Tropical Rain Forest

Questions

1. A biome is considered a desert based on its
   (1) temperature
   (2) winds
   (3) communities
   (4) rainfall

2. Which best describes the area in which you live?
   (1) grassland
   (2) temperate deciduous forest
   (3) taiga
   (4) desert

3. Which one of the following is **not** a biome?
   (1) tundra
   (2) tropical rain forest
   (3) mountain
   (4) savannah

4. Coniferous trees can be found in a
   (1) taiga
   (2) tundra
   (3) temperate deciduous forest
   (4) tropical rain forest

5. A biome with permanently frozen soil and short winters and which contains no trees is called
   (1) a taiga
   (2) a grassland
   (3) a desert
   (4) the tundra

Thinking and Analyzing

1. Although plants in the warm, Amazon region of Brazil are green and lush, they grow in poor soil that contains very few nutrients. The soil and the minerals within it are constantly being washed away. What biome describes the Amazon region?

   Base your answers to questions 2 and 3 on the map and the paragraph below.

   The Adirondack Mountains in upstate New York's Adirondack Park are home to a large variety of trees. Among them are spruce trees and fir trees, which are needle-leaf evergreens. There are also deciduous trees such as maples and oaks. The region is considered a transition between two biomes.

2. What biome is north of the Adirondack Mountains, and what biome is south of the Adirondack Mountains?

3. Explain why the biomes change as you go north.

4. How does elevation affect the types of trees you find in the mountains?

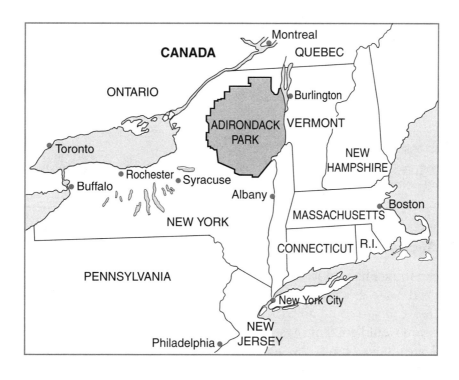

What Are the Patterns of Environmental Change?

Objectives

Describe ecological succession.

Distinguish between primary succession and secondary succession.

Explain what is meant by a pioneer species.

Terms

ecological succession (suhk-SEH-shun): the natural process by which one community of living things is replaced by another community, until a stable climax community appears

pioneer (pih-oh-NEAR) **species:** the first organisms to appear in a barren environment

climax (KLI-macks) **community:** the final community that appears and is not replaced through ecological succession

secondary succession: the process of ecological succession that begins from preexisting soil

primary succession: the process of ecological succession that begins from barren rock

aquatic (uh-KWAH-tic) **succession:** ecological succession that begins with a lake or pond community

---

Ecological Succession

The Adirondack Mountains are located in northern New York State, as shown in the map in Figure 8.3-1. If you could go back in time 200 years, you would see that most of the Adirondack area looked much the same as it does today. It contained the same types of plants, and many of the same animals. Other environments, however, such as the area around Mount St. Helens in the state of Washington, are changing gradually over time in a predictable way. Environments change through a series of events called **ecological succession**.

Ecological succession begins with an area that is *barren*. This means that there is little or no life present in that area. The eruption of a volcano, such as Mount St. Helens, often creates a large area of bare rock. Small organisms, called *lichens* (LIE-kens), are able to survive on the rock. Lichens are the first organisms able to survive in this barren environment. (See Chapter 9, page 342.) The first organisms to appear in a barren environment are called **pioneer species**.

Figure 8.3-1. A map of New York State showing the location of Adirondack Park.

Improving the Soil

Over time, lichens soften the rocks. The rocks break down and eventually form soil that is better suited to support other organisms. The next group of organisms that appear after lichens includes mosses, worms, and insects. The waste products and decayed remains of these organisms continue to enrich and thicken the soil. Small plants, such as ferns and grasses, begin to appear. These plants become homes or food for more insects and small animals. Eventually, these plants die and enrich and thicken the soil more. When the soil is thick enough, and water supplies are adequate, shrubs and finally trees begin to appear. Each new community changes the environment, making it more suitable for the next community. (We discuss and define communities in Lesson 9.1.) Finally, a

community emerges that is not replaced. This community is called the **climax community**. Figure 8.3-2 on the next page illustrates the ecological succession of a barren area into a forest.

Climax Communities

The climax community on Mount St. Helens included the forest of spruce and fir trees and the animals that lived there. This biome is known as a coniferous forest, or taiga. (See Table 8.2-1 on page 309 in the previous lesson.) Through ecological succession, the same type of forest will eventually return to the land around Mount St. Helens.

Different environmental factors, such as climate and water, determine the course of succession in the biome that will eventually form. The colder climate near Mount St. Helens results in a coniferous forest of evergreen trees. The warmer climate in New

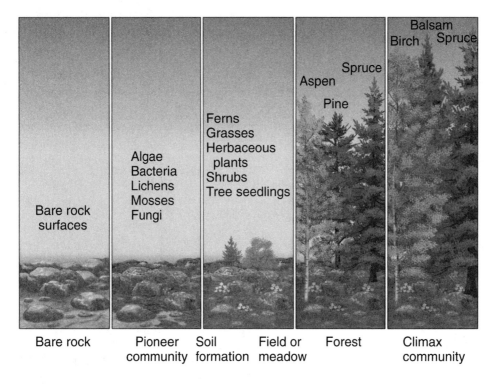

Bare rock surfaces | Algae Bacteria Lichens Mosses Fungi | Ferns Grasses Herbaceous plants Shrubs Tree seedlings | Aspen Pine Spruce | Birch Balsam Spruce

Bare rock | Pioneer community | Soil formation | Field or meadow | Forest | Climax community

Figure 8.3-2.   Ecological succession.

York City results in a deciduous forest of beech and maple trees.

Primary and Secondary Succession

Like a volcano, a forest fire is a natural event that may also destroy an ecosystem. Unlike a volcanic eruption, a forest fire does not destroy the existing soil. Forest fires enrich the soil with minerals from decaying plants and animals that have died in the fire. The process of succession occurs much more rapidly when it begins with fertile soil. Other events, such as floods, windstorms, and human activities can also damage or destroy an ecosystem without removing all of the soil. Succession that begins with the soil already in place is called **secondary succession**. Succession that begins from barren rock, as in Mount St. Helens, is called **primary succession**.

Changing a Lake into a Forest

Another example of succession is **aquatic succession**. Aquatic succession occurs when a lake community becomes a forest community. This happens when sediments are washed into a lake and, over time, fill it in. As the lake becomes shallower, plants and fish that prefer shallow water replace the previous community. Eventually the lake dries up. First, water plants appear, and then land plants establish themselves in the dry lakebed. (See Figure 8.3-3.) All types of succession change an ecosystem in a predictable way and eventually form a climax community.

(a) Pond

(b) Swamp

(c) Meadow

(d) Forest

Figure 8.3-3. Aquatic succession.

Activities

1. Visit one of the botanical gardens or parks in your area. Collect leaves from several different types of trees. Use the Internet or a tree guide to identify these trees.

2. The climax community in New York contains maple and beech trees. In your area, see if you can identify at least one of each of these trees.

Questions

1. As the population of old shrubs decreases in a changing ecosystem, the population of new trees increases. The old community
   (1) destroys the ecosystem
   (2) prepares the ecosystem for the new community
   (3) is the climax community
   (4) is the pioneer community

2. When the glaciers that covered New York State ten thousand years ago melted away, they left only barren rock behind. Eventually, forests formed in the area that was once covered by ice through a process called
   (1) primary succession
   (2) secondary succession
   (3) symbiosis
   (4) conservation

3. Using the diagrams at the right, the correct order of the stages of succession is.
   (1) B→A→D→C
   (2) A→D→C→B
   (3) C→B→A→D
   (4) D→A→C→B

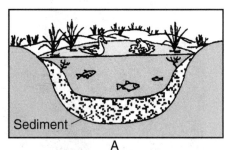
A

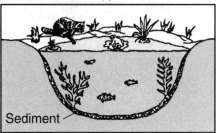
B

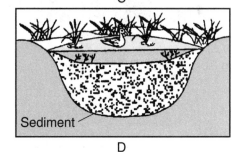
C

D

Questions (continued)

4. A natural disaster completely destroys a forest environment. The first new plants to appear in the recovering environment are
   (1) pine trees
   (2) maple trees
   (3) small shrubs
   (4) mosses and grasses

5. A pioneer community in primary succession most likely contains
   (1) small shrubs
   (2) mature forests
   (3) lichens
   (4) worms and insects

Thinking and Analyzing

1. The flooding from hurricane Katrina destroyed or badly damaged some of the ecosystems of southern Mississippi. What type of ecological succession is likely to take place there? Explain your answer.

2. New York City still has some vacant lots where buildings were taken down. If such a lot is left undisturbed, how do you imagine it will change in the next 50 years? What process will cause the changes you describe?

8.4 How Is Our Climate Changing?

Objectives

Identify factors than can cause a change in global temperatures.

Discuss the possible impact of global warming on our environment.

Terms

global warming: a rise in worldwide temperatures

fossil fuels: an energy source that is formed from the remains of plants or animals

greenhouse effect: a heating of the atmosphere due to gases that trap and hold heat

Global Warming

Have your parents or grandparents ever told you that the winters were much colder when they were kids? Is it possible that the winters are getting warmer? Most scientists think that they are. This increase in the average worldwide temperature is called **global warming**. Scientists believe that Earth's climates are getting warmer over time. The ten warmest years ever recorded happened after 1990. Scientists estimate that the average global surface temperature has increased about 1°F in the past century.

Causes of Global Warming

What is causing global warming? Most scientists who study climate believe that human activity is partially responsible. **Fossil fuels** are materials we burn to create energy. They are made from the remains of plants and animals that died a long time ago. When we burn fossil fuels such as coal, wood, gasoline, natural gas, or oil, we add carbon dioxide and other gases to the atmosphere.

Carbon dioxide and these other gases, such as nitrous oxide and methane, cause the atmosphere to trap and hold more heat. (See Lesson 4.2.) These gases behave like the glass in a greenhouse, allowing sunlight to enter but trapping the heat inside. (See Figure 8.4-1.) The resulting build-up of heat is called the **greenhouse effect**, and the gases that cause it are called greenhouse gases.

Environmentalists hope to slow down global warming by reducing the amount of greenhouse gases put into the atmosphere. They want to reduce worldwide use of fossil fuels.

Not all scientists are convinced that human behavior is a major cause of global warming. There are some scientists who

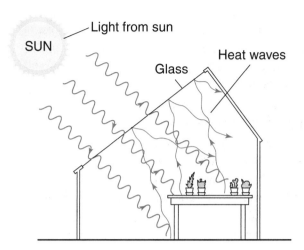

Figure 8.4-1 Light passes through the glass windows of a greenhouse and gets changed into heat. The heat is trapped in the greenhouse.

believe that global warming is just part of a natural cycle of warming and cooling that has been occurring for millions of years. In the past, Earth has been both warmer and cooler than it is now. There have been periods in which ice has covered much more of Earth than it does now. These *ice ages* have come and gone. The last one reached its peak about 18,000 years ago. (See Figure 8.4-2.) Scientists are not sure what causes these natural cycles. Scientists who believe

that this natural cycle is the major cause of global warming feel that changes in our behavior will have little or no effect on Earth's changing temperatures.

Dangers of Global Warming

What could be wrong with a warmer winter? Growing seasons would be longer and more food could be produced. The major problem with global warming starts in the high latitudes near the poles. In this area, the land is covered with huge mountains of ice. As global temperatures have risen, some of this ice has already begun to melt. The melted ice enters the oceans, causing a rise in sea level. If the sea level continues to rise, coastal areas could get flooded. In a worst-case scenario, New York City and Los Angeles would be completely under water.

You may not enjoy cold weather, but some plants and animals require it. Warming temperatures and melting ice are threatening the survival of polar bears and other northern species.

An increase in global temperature can also cause a change in ocean currents. This could significantly affect Earth's climates and weather patterns. Some areas could get more

Figure 8.4-2. The estimated average temperature of Earth has been warmer and cooler many times in the past million years.

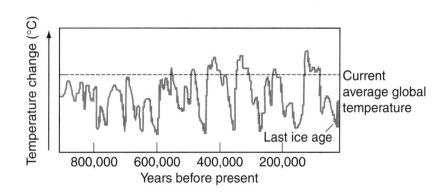

Interesting Facts About Global Warming

Global warming is often in the news. People argue about how bad global warming will get, when it will happen, and what to do about it. If you look at old magazines and newspapers, however, you might be surprised at what they say about Earth's climate. In 1975, many scientists were concerned about global cooling. Magazine articles appeared with the title, "The Coming Ice Age." How could scientists have been so wrong just 30 years ago? Predicting how the climate will change is very difficult. Volcanic eruptions, changes in the amount of energy coming from the sun, and human activities all affect the world's temperatures. Even today, although most scientists are convinced that global warming is happening, there are a few who are not so sure.

rain than in the past, while others could get less. These changes would result in floods in some areas, and droughts in others. Warmer oceans may produce stronger storms, like hurricane Katrina, which struck New Orleans in August of 2005.

Activities

1. Visit the website *http://www.epa.gov/globalwarming/kids/version2.html* and watch the global warming movie. Read each scene carefully. When you are done, take the quiz at the end.

2. Visit the site *http://www.abcbirds.org/climatechange/NewYork.pdf.* You will find a lot of information about how global warming might affect birds in New York City.

a) Name one bird that might become less common, and one that might become more common, if temperatures continue to rise.

b) Search the Internet for photographs or drawings of these birds. Print them out and bring them to class.

Questions

1. Scientists believe that human beings contribute to global warming because we
   (1) are warm-blooded animals
   (2) build dams that decrease the flow of rivers
   (3) burn fuels that produce carbon dioxide
   (4) consume too much of Earth's water

2. Which of the following is a danger that may result from global warming?
   (1) food production can increase
   (2) the oceans can freeze
   (3) ice will melt causing floods
   (4) the excess heat will cause forest fires

3. In the past, Earth's climate was
   (1) always warmer than it is now
   (2) always cooler than it is now
   (3) exactly the same as it is now
   (4) sometimes warmer and sometimes cooler than it is now

Thinking and Analyzing

1. The release of carbon dioxide into the air may be partly responsible for the increase in worldwide temperatures. State one way that people in New York can decrease the amount of carbon dioxide that we are releasing into the air.

2. Look back at Figure 8.1-5 on page 302. Some scientists predict that by the year 2050 there will be no snow at all on Mount Kilimanjaro. Give one reason why scientists feel that the snow is disappearing from the mountain.

3. When you enter a car on a sunny day in the winter, the car is warm even though the temperature outside is not. Explain why this occurs.

Review Questions

Term Identification

Each question below shows two terms from Chapter 8. One of the terms is defined.
(1) Choose the term that matches the definition.
(2) Describe how the two terms are different. Following each term is the section (in parenthesis) where the description or definition of that term is found.

1. *Elevation (8.1) — Latitude (8.1)*
   A measure of distance from the equator

2. *Temperate (8.1) — Tropical (8.1)*
   A climate having cold winters and warm summers

3. *Taiga (8.2) — Tundra (8.2)*
   A treeless biome with soil that is always frozen

4. *Deciduous temperate forest (8.2) — Tropical rain forest (8.2)*
   A biome with a warm, wet climate and a climax community containing broadleaf evergreens

5. *Desert (8.2) — Savannah (8.2)*
   A tropical grassland

6. *Pioneer species (8.3) — Climax community (8.3)*
   The first organisms to appear in a barren environment

7. *Global warming (8.4) — Ecological succession (8.3)*
   The natural process by which one community of living things is replaced by another community

Multiple Choice (Part 1)

1. The diagrams below show the plant communities present in the same area at different times over a 200-year period following a forest fire.

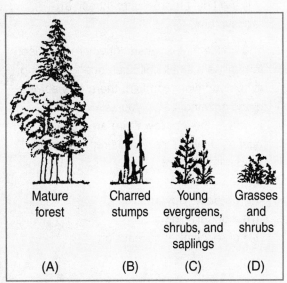

What is the correct sequence of these plant communities following the forest fire?
(1) B→A→D→C   (3) B→D→C→A
(2) B→C→D→A   (4) B→A→C→D

2. Which factor contributes to global warming?
   (1) increased use of solar-powered cars
   (2) increased burning of fossil fuels
   (3) better long-term weather forecasts
   (4) changing distance between Earth and the sun

3. Which description best describes a biome?
   (1) a species of organisms
   (2) all the organisms in an ecosystem
   (3) the nonliving factors of the environment
   (4) a region of Earth with a particular climate

4. A climax community is
   (1) the first species to appear in an ecosystem
   (2) a species that depends on animals for food
   (3) a species that returns after a natural disaster
   (4) the final organisms to inhabit an ecosystem

5. You notice that a vacant lot where you used to play as a small child has been changing. When you were younger, there was some grass growing. Later, you saw weeds covering the lot. Now there are trees and bushes growing there. This replacement represents part of
   (1) a food chain
   (2) an energy pyramid
   (3) a desert biome
   (4) an ecological succession

6. We would expect to find the warmest climates at
   (1) high latitudes and high elevations
   (2) low latitudes and low elevations
   (3) high latitudes and low elevations
   (4) low latitudes and high elevations

7. The main factor that results in a desert biome is
   (1) low amounts of rainfall
   (2) sandy soil
   (3) high temperatures all year
   (4) hot summers and cold winters

8. What would you expect to find in a deciduous forest biome?
   (1) trees that remain green all year
   (2) soil that is frozen all year
   (3) trees that lose their leaves in the fall
   (4) small shrubs and cactus

9. How might you expect the biomes to change as you climb a high mountain?
   (1) taiga→deciduous forest→tundra
   (2) deciduous forest→taiga→tundra
   (3) taiga→tundra→deciduous forest
   (4) deciduous forest→tundra→taiga

10. Scientists are worried about melting ice in Greenland because the melting could lead to
    (1) more severe hurricanes
    (2) colder weather
    (3) flooding in coastal areas
    (4) greater rainfall

Thinking and Analyzing (Part 2)

1. The Galapagos Islands are a chain of small volcanic islands located in the Pacific Ocean near the equator. Some of these islands were formed a long time ago, while others were formed more recently. A photograph of one of the Galapagos Islands is shown above. Examine the photograph and determine if it was taken on a young or old island. Explain your answer.

2. Santa Cruz and Fernandina are both part of the Galapagos Islands. However, Santa Cruz is an island covered with thick forests, while Fernandina Island has only cactus and mangroves growing on barren rock. Predict what the environment on Fernandina might look like in the future.

3. Explain the difference between a pioneer organism and a climax organism.

4. People travel from all over the world to see autumn in New York State or New England. Indicate the name of the biome they are visiting. Explain what is unique about autumn in this biome.

5. Read the paragraph below and answer the question that follows it.

   Trees and other green plants take in carbon dioxide and release oxygen during the process of photosynthesis. During photosynthesis, green plants act as a filter that absorbs carbon dioxide from the atmosphere. When trees are cut down, they stop removing carbon dioxide from the atmosphere. When trees are burned, they release carbon dioxide into the atmosphere. Some scientists feel that there is a direct link between the cutting down of the tropical rain forest and global warming.

   Argue that the destruction of the tropical rain forest affects global warming, and give a reason to support your position.

Chapter Puzzle (*Hint:* The words in the puzzle are terms used in the chapter.)

Across

1 a climate with warm summers and cold winters

3 the final community that appears due to ecological succession is called the _____ community

5 a region that is characterized by a particular climate, soil, and community of plants

7 distance from the equator, expressed in degrees

9 the average weather in an area

10 a tropical grassland

12 a biome that receives less than 10 inches of rain per year

13 a cold-climate biome that contains needle-leaf evergreen trees

15 the first organisms to appear in a barren environment are called the _____ species

16 an imaginary line drawn around the center of Earth

17 energy sources formed from the remains of plants and animals are called _____ fuels

Down

1 a hot and moist climate

2 a biome found in warm, wet climates, containing broadleaf evergreen trees

4 when heat is trapped by carbon dioxide, it produces a _____ effect

6 height above sea level

8 a treeless biome with frozen soil

11 ecological succession in a pond community is called _____ succession

12 trees that drop their leaves in winter are _____ trees

14 the rise in worldwide temperatures is called _____ warming

Chapter 9

Ecosystems and Interdependence

Contents

An underwater ecosystem.

What Is This Chapter About?

The success of an ecosystem depends on the relationships among all the organisms living in it.

In this chapter you will learn:

1. The organisms living in an ecosystem make up a community.

2. The population of a species may change as environmental factors change.

3. The population of one species affects the populations of other species in the community.

4. Organisms depend on each other to survive.

5. Environmental changes can be sudden or gradual.

Career Planning: Environmental Scientist

Environmental scientists are scientists who monitor and protect the environment. After the collapse of the World Trade Center in 2001, scientists from the Environmental Protection Agency (EPA) were called in to study the quality of the air in the surrounding buildings. In New Orleans in 2005, much of the city was flooded after hurricane Katrina. The EPA was asked to determine which homes were safe enough for their owners to return to. Environmental scientists are responsible for determining the quality of our soil, air, and water. The EPA, created by President Nixon in 1970, makes sure that people, corporations, and the government are taking care of the environment.

Internet Sites:

http://www.factmonster.com/ipka/A0776202 .html An introduction to the way that certain organisms depend on each other.

http://www.epa.gov/sunwise/kids/kids_ozone .html Read about the destruction of the ozone layer.

http://library.thinkquest.org/C005824/extinction .html How did the dinosaurs become extinct? Here are several theories.

9.1 What Are the Parts of an Ecosystem?

Objectives

Distinguish between biotic and abiotic factors in an ecosystem.

Distinguish between a population and a community.

Terms

biotic (by-AH-tik) **factors:** living parts of an environment

abiotic (AY-by-AH-tik) **factors:** nonliving parts of an environment

species: a group of organisms of the same kind that produce offspring that can reproduce

population (pa-pew-LAY-shun): all the members of a particular species that live within a given area

community: all the different populations that live within a given area

---

What Are the Living and Nonliving Parts of an Ecosystem?

You have already learned in Lesson 7.1 that an environment consists of both living and nonliving things. Scientists use the term **biotic factors** to refer to the living parts of the environment. **Abiotic factors** are the nonliving components of an environment, such as water, air, soil, and temperature. An *ecosystem* is an area where biotic and abiotic factors interact.

Species

An ecosystem contains many different types of organisms. A particular type of organism is called a **species**. Organisms are considered members of the same species if they can produce offspring who can also reproduce.

For example, poodles, collies, beagles, and German shepherds are all dogs. Although they look somewhat different, they can mate with any other type of dog and produce puppies. When these puppies grow up, they will also be able to mate successfully with any other type of dog.

A horse and a donkey are not members of the same species. When a horse mates with a donkey, the offspring are called mules. Mules, however, cannot produce offspring. In nature, animals of different species seldom try to mate.

Communities

If you were asked to count the population of your community, you would probably include only one species—human beings.

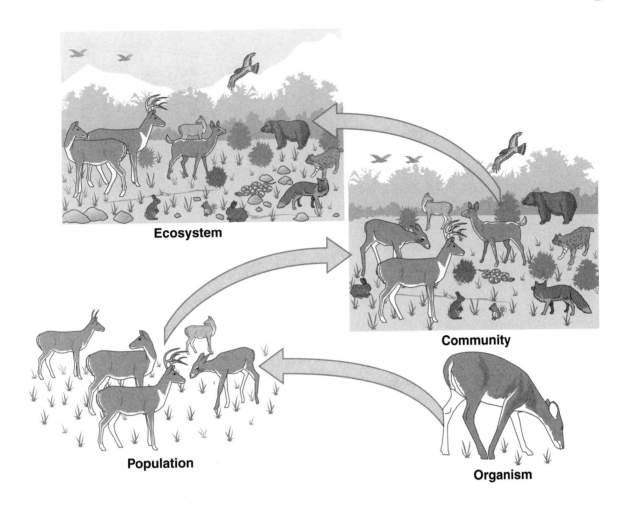

Figure 9.1-1. Parts of an ecosystem.

But your community also contains a population of dogs, cats, and other living things. A **population** is defined as all of the members of one particular species that live in the same area. An ecosystem typically contains many populations. A **community** consists of *all* of the populations within an ecosystem. When we refer to a community, we are referring to *all* the living things in a given area. (See Figure 9.1-1.)

Figure 9.1-2 on page 330 shows a small pond community. What are the populations that make up this community? The mallard ducks, bullfrogs, and trout are obvious populations in this community. Each species of plant is part of its own population. Although they are not shown in the diagram, a pond community also contains populations of bacteria and plankton and other living things.

Different types of communities contain different populations. The populations found in a desert community are very different from the populations found in a

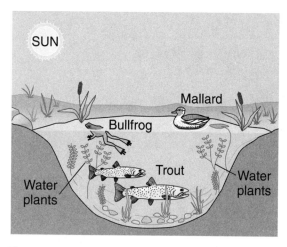

Figure 9.1-2. A pond community.

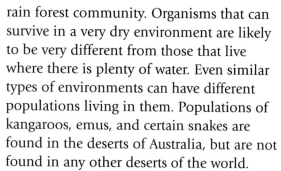

Figure 9.1-3. A small ecosystem.

rain forest community. Organisms that can survive in a very dry environment are likely to be very different from those that live where there is plenty of water. Even similar types of environments can have different populations living in them. Populations of kangaroos, emus, and certain snakes are found in the deserts of Australia, but are not found in any other deserts of the world.

Scientists study the populations in an ecosystem to keep track of the health of that ecosystem. It is normal for populations to change over time. But a sudden decrease or increase in one population might be a warning sign that some change in the

environment is harming the community. To determine if the ocean is safe for swimming, environmental scientists measure the amount of bacteria called *E. coli* in the water. If the level of *E. coli* is too high, the water is not safe.

An aquarium is a small ecosystem (see Lesson 7.2). In setting up an aquarium, you may include several populations of plants and animals. You must also provide the necessary abiotic factors such as light, heat, gravel, water, and oxygen. If your ecosystem is successful, the community will thrive. The population of each species will be maintained. (See Figure 9.1-3.)

Activity

Although a tree may be a member of a population, it may also be an ecosystem for smaller plants and animals. Select a tree in your neighborhood and look for other organisms that live on or in that tree. List all the different populations that you find. Remember to carefully examine the entire tree. Look high and low and for large and small organisms . . . very small organisms.

Questions

Use the figure below, which shows a pond ecosystem, to answer questions 1-4.

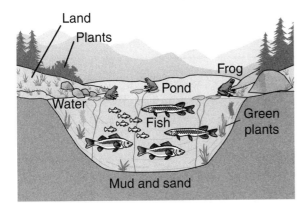

1. Which correctly describes a population in this community?
   (1) 11 fish   (3) 14 animals
   (2) 3 frogs   (4) 17 organisms

2. Within this ecosystem, the pond community consists of
   (1) 11 fish
   (2) 11 fish and 3 frogs
   (3) 11 fish, 3 frogs, and several plants
   (4) 11 fish, 3 frogs, several plants, mud, sand, and water

3. Which is considered a biotic factor in this ecosystem?
   (1) air   (3) water
   (2) plants   (4) sunlight

4. How many different populations of animals are shown?
   (1) 1   (3) 17
   (2) 11   (4) 4

5. Which term includes all of the others?
   (1) community   (3) organism
   (2) population   (4) species

6. Which pair of organisms belongs to the same species?
   (1) blue jay and hawk
   (2) shark and salmon
   (3) lion and lioness
   (4) spider and fly

Thinking and Analyzing

1. Place the following terms in order, from smallest to largest: ecosystem, population, community, organism.

2. Identify at least four populations that live in your neighborhood. Choose at least two populations that are dependent on each other, and explain how these two populations interact.

9.2   What Determines the Size of a Population?

Objectives

Define carrying capacity.

Determine the carrying capacity of an ecosystem for a population based on a graph of population over time.

Identify some limiting factors that determine carrying capacity.

Terms

limiting factors: conditions that limit the size of a population within an environment

carrying capacity: the number of individuals of a given species that can be supported by an ecosystem

---

Why Do Populations Stop Growing?

Look at the graph in Figure 9.2-1. It shows the change in population of New York City since 1880.

Notice that between 1890 and 1950 the population increased steadily, from around 1 million to nearly 8 million. What has happened since 1950? The population has gone down a little, and then up a little, but overall there has been little change.

In a stable ecosystem, populations show the same kind of pattern. Once they reach a certain number, they tend to level off, or vary within a certain range. Why? Read on.

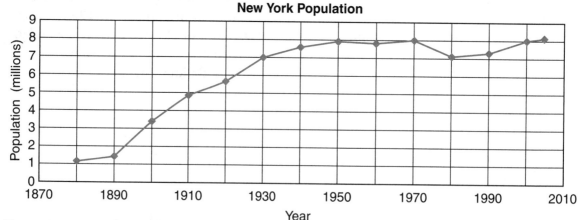

Figure 9.2-1.   New York City's changing population.

Limiting Factors

If the population of New York continued to grow, the city would eventually run out of space. There would not be enough water, housing, adequate transportation, or schools to support these additional people. Conditions that limit the population an area can support are called **limiting factors**. Let us examine the limiting factors in a typical ecosystem.

Look back at the food web shown in Figure 7.5-1 on page 276. What is a factor limiting the population of rabbits? We can see that one factor is the amount of grasses and shrubs the rabbits need for food. Another factor is the number of grasshoppers. Rabbits and grasshoppers compete for the same food supply. An increase in the population of grasshoppers decreases the amount of food available for the rabbits. Other limiting factors for rabbits include a lack of water and shelter.

A shortage of grasses and shrubs does not limit just the population of the primary consumers that eat them. The secondary consumers depend on them as well. A loss of plants means a loss of rabbits and small birds. A loss of rabbits and small birds means less food for hawks and foxes as we go up the food chain. (See Lesson 7.7)

Carrying Capacity

The number of individuals of a given species that can survive in an ecosystem is called the **carrying capacity** of that ecosystem. In an ecosystem, different organisms have different carrying capacities. For example, the carrying capacity for foxes and hawks is usually smaller than it is for rabbits and small birds, because large animals need more food and water than small animals. Therefore, you would expect to find fewer foxes and hawks and more rabbits and small birds.

When a small population of a new organism enters an ecosystem, it often grows very rapidly. At first, its numbers are much smaller than the carrying capacity. As the population increases, it is eventually affected by one or more limiting factors. The availability of food, water, or shelter can limit further growth. The result is shown in Figure 9.2-2.

Figure 9.2-2. Change in a population over time.

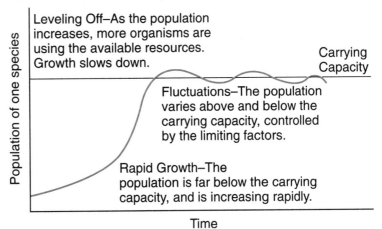

Population Growth

Leveling Off–As the population increases, more organisms are using the available resources. Growth slows down.

Carrying Capacity

Fluctuations–The population varies above and below the carrying capacity, controlled by the limiting factors.

Rapid Growth–The population is far below the carrying capacity, and is increasing rapidly.

Population of one species

Time

Fluctuations in Population

As shown in Figure 9.2-2, the population can exceed the carrying capacity. When this occurs, there are not enough resources to support that population. Where there is *over*population, some organisms will not be able to find enough food and will die. This results in a decrease in the population. This smaller population eats less food, so the food supply increases. An increase in the food supply results in an increase in the population. These up-and-down changes are called *fluctuations* (FLUHK-choo-WAY-shuns).

The carrying capacities of all the animals in an ecosystem are interrelated. Figure 9.2-3 shows what would happen to a prey population of zebras if a population of lions were introduced to the area. At first, as the number of lions increases, the number of zebras decreases. Eventually, there is not enough prey for the lions to feed on. The lion population begins to decrease. With fewer lions eating them, more of the zebras survive. The prey population, the zebras, begins to increase again. This cycle of population change repeats.

Competition

Over time, the population of lions stabilizes. It might change slightly from year to year,

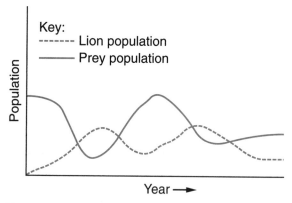

Figure 9.2-3. Populations of predator and prey.

but it is always limited by the amount of prey available. What happens if a population of hyenas that hunts the same prey as the lions enters the ecosystem? Less food becomes available for the lions, and their population probably decreases. In an ecosystem, the carrying capacity for a particular animal is limited by the competition it gets from other animals.

The human population in New York City has remained at about 8 million people for thirty-five years. Perhaps a population of 8 million people is the city's carrying capacity. *Stop and Think:* What factors can you think of that limit the carrying capacity of people in New York City? *Answer:* The availability of affordable housing; the availability of jobs.

SKILL EXERCISE—*Graphing*

Chickadees are small songbirds that eat seeds. They are common in New York City during the winter months. Chickadees are attracted to feeding stations containing sunflower seeds. They sometimes eat seeds right out of your hand. In recent years, more and more people have taken up the hobby of setting up and maintaining bird feeders.

The data to the right shows the number of chickadees found each year in a New York community during an annual event called the winter count.

1. Use the data to draw a line graph on a piece of graph paper. Label the *x*-axis "Year" and the *y*-axis "Number of chickadees." (For help, see the graph below.)
2. Make up an appropriate title for this graph.
3. What was the apparent carrying capacity of the community for chickadees between 1991 and 1995? In 2003?
4. What is one possible reason for this change?

Year	Number of chickadees found in winter count
1991	99
1992	95
1993	101
1994	102
1995	100
1996	120
1997	133
1998	145
1999	160
2000	162
2001	159
2002	157
2003	160

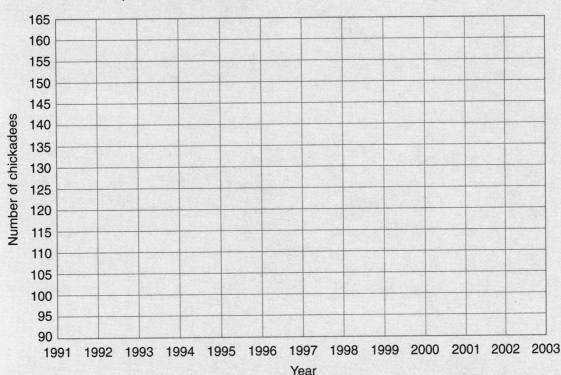

Interesting Facts About Human Population Growth

In the 1850s, Thomas Malthus, a British economist, predicted that the rate of human population growth would result in mass starvation, as the number of people outgrew the carrying capacity of the global environment. Human population did grow, even more than Malthus predicted, yet there was no mass starvation. Advances in science and technology have improved our ability to produce food. The carrying capacity of Earth today is much larger than Malthus predicted it would be. For now, at least, science has given us a way to produce enough food and water for Earth's human population. However, although our food and water supply is theoretically large enough to support Earth's human population, it is not evenly distributed. Starvation is a serious problem in some parts of the world due to our inability to distribute the food, not our inability to grow it.

Questions

1. Which of the following factors might cause a decrease in the population of hawks in an ecosystem?
   (1) an increase in the availability of prey
   (2) an increase in the population of a competing predator
   (3) an increase in the water supply
   (4) an increase in the plant population

2. The maximum population that an ecosystem can support is called its
   (1) limiting factor
   (2) carrying capacity
   (3) community
   (4) habitat

3. How does the size of a population change as it reaches its carrying capacity?
   (1) It increases rapidly.
   (3) It decreases slowly.
   (2) It decreases rapidly.
   (4) It levels off, showing only small changes.

Thinking and Analyzing

Use the graph at right, which shows the population of gazelles in a grassland ecosystem, to answer questions 1-4.

1. What does line *A* represent?
2. Between what years did the gazelles grow rapidly?
3. What might have caused the change in the gazelle population that began in late 2001?
4. How did the gazelle population change between 1998 and 2002? Explain why this change occurred.

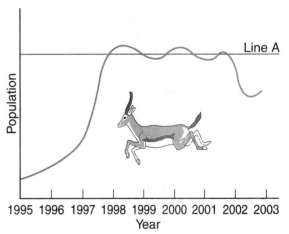

Gazelle Population from 1995 through 2003

9.3 How Do Organisms Depend on Other Organisms?

Objectives
To be able to tell the difference between symbiosis and other relationships that organisms have, such as predator-prey and competition.

Identify three types of symbiotic relationships.

Terms
symbiosis (sim-by-OH-sis): a relationship between different species living in close, permanent association

mutualism (MEW-chew-uhl-lis-uhm): a symbiotic relationship in which both species benefit

commensalism (koh-MEN-suh-liz-uhm): a symbiotic relationship in which one species benefits, while the other is not affected

parasitism (PAH-ruh-sih-tiz-uhm): a symbiotic relationship in which one species benefits, while the other is harmed

What Interactions Take Place in an Ecosystem?

You have already learned how animals depend on their environment to survive. The environment provides water, air, sunlight, shelter, and food. Animals obtain food by eating plants or other animals. Food provides the energy necessary for survival. In a predator-prey relationship, one animal feeds on the other. The relationship between plants and herbivores is similar to a predator-prey relationship, because herbivores eat plants.

Another relationship between organisms is competition. Although organisms in competition do not rely on each other for food, their lives are closely connected. These organisms rely on the same food source. If one population grows, the other might shrink, because the food supply is reduced.

Symbiosis

A number of species relate to each other in yet a different way. In these relationships, one organism can get food, shelter, or even transportation from another organism. These organisms live closely together, and at least one of them always benefits. This permanent relationship is called a symbiotic relationship, or **symbiosis**. The word "symbiosis" actually means, "living together." Scientists recognize three different types of symbiotic relationships: **mutualism**, **commensalism**, and **parasitism**.

Mutualism

Mutualism is a symbiotic relationship in which both species benefit. For example, the relationship between a flower and a butterfly benefits both of them. Many flowers can reproduce only when the pollen from one flower is transferred to another flower. This process is called *pollination*. Flowers rely on certain insects, such as butterflies and bees, to carry the pollen from flower to flower. The insect picks up pollen when it feeds on the nectar in the flower. The insect gets food from the flower, and the flower gets pollinated. Figure 9.3-1 shows a butterfly feeding on the nectar of a flower. At the same time, pollen sticks to the butterfly and gets transported to other flowers.

Did you know that termites eat wood but cannot digest it? They lack the chemical necessary to break down cellulose, a substance found in wood. Certain microscopic protists, called protozoa (see Lesson 6.4), live inside the gut of the termite and break down the cellulose into sugar. The termites get sugar from the protozoa, and

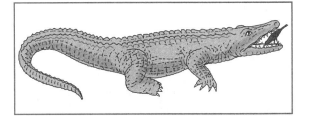

Figure 9.3-2. Another example of mutualism. The Egyptian plover gets food, while the crocodile gets its teeth cleaned.

the protozoa get the ground-up wood and a moist environment from the termites. Both organisms benefit from the nutrition provided by the sugar. This is another example of mutualism.

Figure 9.3-2 shows a bird, called an Egyptian plover, in the mouth of a crocodile. Is the crocodile about to eat the bird? The crocodile does not eat it, because the bird serves a useful purpose. The bird is feeding on leeches and other scraps of food in the crocodile's mouth. Both the crocodile and the plover benefit from this relationship. The bird gets food while the crocodile gets its teeth cleaned.

Commensalism

In all symbiotic relationships, one organism benefits. Sometimes the other organism also benefits, as in the case of the Egyptian plover and the crocodile. Other times, one organism benefits and the other organism is not affected at all. It neither benefits from the association nor is harmed by it. This type of symbiotic relationship is called **commensalism**. An example of commensalism is the relationship between

Figure 9.3-1. The relationship between a flower and a butterfly is an example of mutualism.

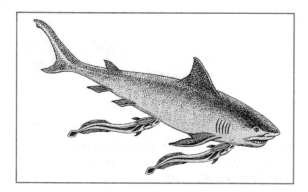

Figure 9.3-3. An example of commensalism. The remora benefits, but the shark does not.

a remora and a shark. Figure 9.3-3 shows how the remora, or suckerfish, sticks to the shark and gets a free ride. When the shark eats, the remora gets the scraps. In this relationship, the remora benefits but the shark does not.

Another example of a relationship between two organisms where one benefits and the other is not affected is that between a humpback whale and the barnacles living on its skin. The barnacles do not affect the whale. The barnacles gather food as they are transported through the water. Figure 9.3-4 shows the barnacles living on the head of a whale.

Figure 9.3-4. Another example of commensalism is the relationship between barnacles and a humpback whale.

Parasitism

In mutualism, and commensalism, neither organism is harmed. A third symbiotic relationship, called **parasitism**, occurs when one organism benefits and the other organism is harmed. Parasitism is discussed fully in the next lesson.

Activity

A coral reef is an ecosystem filled with many symbiotic relationships. Here is a description of some of the organisms living in a coral reef. Read the description and determine some of the symbiotic relationships described.

Look at the coral in the figure on the next page. At first glance, coral does not appear to be a living thing, but it is. Most corals are colonies, or groups of many individual animals that share a common external skeleton. Within the coral are tiny animals called coral polyps. Each coral polyp looks like an upside down jellyfish. Within each polyp live algae. Algae are one-celled plants that undergo photosynthesis. The algae give off oxygen and other nutrients that the coral polyp uses to live. The polyp gives the algae carbon dioxide and other materials necessary for its survival.

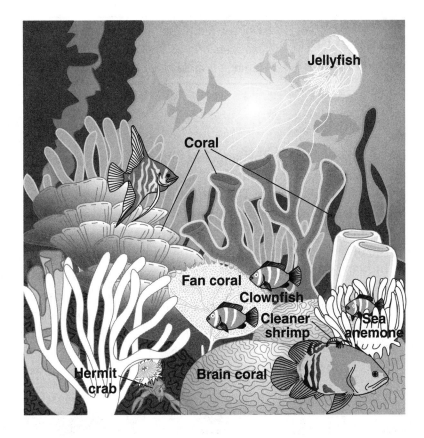

Another interesting animal in the coral reef is the cleaner shrimp. Cleaner shrimp eat ick. Ick is a harmful parasite that grows on the skin of fish. Fish infested with parasites swim to the cleaner shrimp. Cleaner shrimp clean fish by eating the parasites and dead skin.

Clownfish are *immune* to the sting of the sea anemone's tentacles. That means the sting does not hurt them. The clownfish avoids predators by hiding in the tentacles of the sea anemone. The sea anemone is not affected by the presence of the clownfish.

Hermit crabs will pick up sea anemones or sponges and place them on their backs to help them hide from predators. The crabs will be protected, and the sea anemone is not affected.

Identify the type of symbiosis that exists between the organisms listed below. Copy the diagram into your notebook and write your response on the arrow's line.

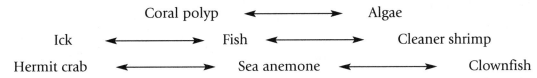

To learn more about the coral reefs, and to test yourself on how much you have learned, try this Web site on the Internet:

http://middle.capemayschools.com/mt4/TheCay/CoralReefs.htm

Interesting Facts About Clownfish

Did you know that all clownfish are born as males? A clownfish stays a male as long as there is a female around. When there is no longer a female, a male clownfish turns into a female.

Questions

1. Bacteria that live in the large intestine of humans obtain their nutrients from the food passing through. These organisms are often called "friendly bacteria," because they help digest our food. The relationship between these bacteria and humans is best described as
   (1) mutualism   (3) parasitism
   (2) competition   (4) commensalism

2. Which term includes the other three?
   (1) mutualism   (3) symbiosis
   (2) commensalism   (4) parasitism

3. Small fish called clownfish live within the tentacles of a sea anemone. The tentacles contain stinging cells that do not harm the clownfish, but keep predators away. The anemone is not affected by the presence of the clownfish. This relationship is best described as
   (1) mutualism   (3) parasitism
   (2) competition   (4) commensalism

4. A close relationship between two organisms in which one benefits and the other is harmed is called
   (1) mutualism   (3) commensalism
   (2) parasitism   (4) cooperation

Thinking and Analyzing

1. The photograph at right shows lichen growing on the branch of a tree. Lichen is actually made up of two organisms, an alga and a fungus. The alga, a one-celled plantlike organism, provides the fungus with food. The alga, which normally lives in water, obtains its water and carbon dioxide from the fungus. What type of relationship exists between the alga and the fungus in lichen? Justify your answer.

2. When asked to describe the three types of symbiosis, a student simply handed in the following three pairs of pictures. In your notebook, label each pair with the correct symbiotic relationship it represents.

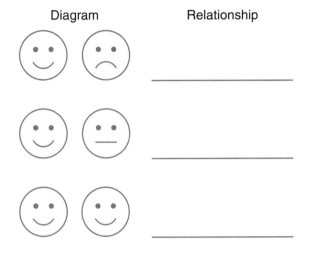

Diagram   Relationship

_____

_____

_____

Questions 3 and 4 are based on the short paragraph below.

Mistletoe is a plant that grows on the woody parts of trees. Mistletoe attaches to the wood of the tree and sucks out water and valuable nutrients. Birds eat the berries of mistletoe and carry the seeds to other trees where more mistletoe plants grow.

3. What is the relationship between trees and mistletoe? Explain your answer.

4. What is the relationship between birds and mistletoe? Explain your answer.

9.4 How Are Organisms Harmed in a Symbiotic Relationship?

Objectives

Identify different types of parasites.

Distinguish between a parasite and a host.

Explain why a parasite does not kill its host.

Terms

parasite (PAH-ruh-site): the organism that benefits from a parasitic relationship

host: the organism that is harmed from a parasitic relationship

brood: a group of birds hatched at the same time and cared for by the same mother

---

Parasites and Hosts

In mutualism and commensalism, one of the organisms always benefits. The other organism in the relationship either benefits (in mutualism), or is not affected (in commensalism). A third type of symbiosis is parasitism. In parasitism, one of the organisms benefits and the other is harmed. A **parasite** is an organism that lives on or in the body of another, usually larger, organism, and gets its food from it. YES, parasitism can be thought of as a predator-prey relationship, with the prey larger than the predator. The organism that is harmed is called the **host**.

There are many types of parasites. They range from very small bacteria to large plants. Sometimes parasitic bacteria enter your body and make you ill.

Parasites Can Suck Blood or Take Nutrients

Leeches that suck blood from a host are parasites as well. Figure 9.4-1 shows parasites such as ticks and fleas that may live on your dog. Your dog may also be a host

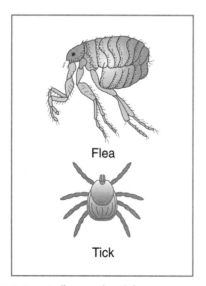

Flea

Tick

Figure 9.4-1. A flea and a tick are two examples of parasites.

for roundworms. Roundworms live inside your dog's small intestines and eat digested food. Parasites can be plants as well. Mistletoe is a plant that lives on the branches of trees. It harms the host tree by taking water and nutrients from it.

Parasites Do Not Kill Their Hosts

How is the relationship between parasite and host different from that of predator and prey? A parasite can be thought of as a small predator eating a large prey. As in a predator-prey relationship, one organism is eating the other. Remember that a parasite lives on or in the body of the host. This is not the case for a predator and its prey. A predator kills its prey. A parasite usually does not kill the host. If the host dies, the parasite has no place to live and needs to find another host.

Parasitic Fungus

Athlete's foot is a skin disease caused by a fungus. This fungus grows in the warm, dark, moist environment between the toes. Athlete's foot fungus is a parasite, because it is living on the skin of its host.

Tapeworm

Another parasite that often affects dogs, cats, and even humans is the tapeworm. A tapeworm, as shown in Figure 9.4-2, is a long, flat organism that lives in the small intestine of the host. The head of the tapeworm attaches itself to the wall of the

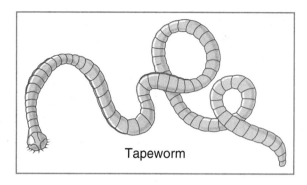
Tapeworm

Figure 9.4-2. A tapeworm is a parasite.

small intestine. The tapeworm's body absorbs nutrients from digested food through its skin. The tapeworm can cause stomach ache, diarrhea, constipation, or weight loss in its host. It produces eggs that come out of the host's rectum. These eggs may spread the infection to other hosts.

Brood Parasites

The brown-headed cowbird is a common bird in New York City. The cowbird does not build nests, incubate eggs, or care for its young. Instead, it lays eggs in the nests of songbirds. Sometimes it even throws the other bird's eggs out of the nest. The mother bird that built the host nest then incubates the eggs and cares for the cowbird hatchlings as if they were her own brood. A **brood** is a group of birds hatched at the same time and cared for by the same mother. The cowbird hatchlings usually get most of the food, and sometimes the other hatchlings die. A cowbird is an example of a *brood parasite*.

Interesting Facts About Leeches

Leeches are parasites that have been used to treat human diseases for thousands of years. It was believed that leeches sucked out the bad blood, along with the disease, thus curing the patient. The leech attaches to the skin using three sharp teeth. It injects a chemical, called an anticoagulant (AN-tie-coh-AG-you-lent), into the skin. Anticoagulants keep blood flowing freely and prevent it from clotting.

Modern medicine has taken a new look at the benefits of leeches. Now, doctors use leeches to increase blood flow in the skin, especially after plastic surgery. One problem faced after surgery is the destruction of veins in the treated area. Damaged veins can't carry blood back to the heart, so blood collects at the wound. The buildup of blood prevents the wound from healing. Leeches help to remove the excess blood, improve circulation, and allow healing to occur.

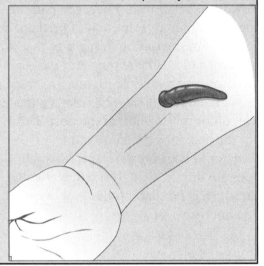

Questions

1. Which statement best describes the symbiotic relationship between a tick and a dog?
   (1) The tick is the host and it benefits.
   (2) The tick is the parasite and is harmed.
   (3) The dog is the host and is harmed.
   (4) The dog is the parasite and is harmed.

2. In parasitism,
   (1) both organisms are harmed
   (2) both organisms benefit
   (3) one organism benefits and the other is not affected
   (4) one organism benefits and the other is harmed

3. A tapeworm is an organism that lives
   (1) inside the body of its host
   (2) on the body of its host
   (3) inside the body of its parasite
   (4) on the body of its parasite

4. All parasites
   (1) feed from their hosts
   (2) kill their hosts
   (3) benefit their hosts
   (4) benefit from their hosts

Thinking and Analyzing

1. Complete the analogy:
   Predator is to prey as parasite is to _____.

2. Compare and contrast parasitism to mutualism and commensalism.

9.5 How Are Organisms Affected by Environmental Change?

Objectives

Distinguish between gradual changes and sudden changes in an environment.
Identify factors than can cause a sudden change in a population.

Terms

extinct (ehks-TEENKT): species that no longer exist

ozone (OH-zone) **layer**: a region in the atmosphere that contains a large concentration of ozone gas, which blocks most ultraviolet radiation from the sun

ultraviolet rays: high-energy radiation from the sun that has been linked to skin cancer

Climates Can Change

If you could go back in time and visit New York City 15,000 years ago, you would probably think you were someplace else. Most of the city would be covered by several hundred feet of ice. During this ice age, ice covered nearly all of New York State. Earth's climate has changed several times in the past, getting warmer, cooler, and then warmer again. Even if cooling began again next year, it would take many thousands of years before ice would cover New York again. So why should we worry?

The particular environment in which an organism lives is called its *habitat*. Small changes in a habitat can greatly affect the organisms living in it. As you learned in Lesson 9.3, a coral reef is an example of a habitat. Coral reefs are found in warm, clear water in many places in the world. Many varieties of fish depend on the coral for food and protection. Recently, scientists have observed a condition called bleaching that is destroying the coral. Coral bleaching is caused by the loss of algae that normally live in the inner cells of the coral. The coral and the algae both benefit from this symbiotic relationship. The algae provide nutrients that help the coral grow and build reefs, and the coral provides shelter and waste products that the algae use for photosynthesis. A change in the water temperature of only 2 degrees Celsius may be largely responsible for coral bleaching. If the coral reefs disappear, the organisms that depend on them may disappear as well.

Extinction

When all the members of a species die out, that species is said to be **extinct**. That organism will never again exist. A temperature change of 2 degrees seems very small to us, yet in some ecosystems, even a change this small can destroy the habitat and eliminate the community.

Destruction of Habitat

An environmental change in temperature is not the only factor that affects the survival of living things. Some environmental changes are the result of human activity. When forests are cleared to provide space for farms or homes, the organisms that depend on the trees for survival are threatened. Sometimes these organisms move to another area that can support them. Unfortunately, not all species are able to find other places to live. The ivory-billed woodpecker, shown in Figure 9.5-1, is now close to extinction because of the destruction of its habitat, the southern hardwood forests.

Organisms Adapt to Gradual Changes

Climates change over long periods of time. Many species are able to adjust to these

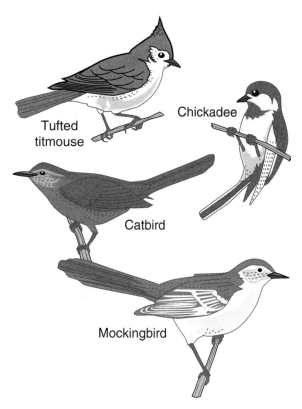

FIGURE 9.5-2. Some species of birds that are found in New York City.

changes. For example, as winters in New York City have become warmer, some birds have become more numerous. Among these are the mockingbird and the tufted titmouse. Other birds, such as the catbird and the chickadee, are likely to become less common in New York if the climate continues to warm. However, none of these species is threatened with extinction. Gradual changes in the environment permit many species to adjust or move to a more favorable location. (See Figure 9.5-2.)

Sudden Changes

Sometimes an environment changes so suddenly and drastically that the ecosystems

FIGURE 9.5-1. The ivory-billed woodpecker is close to extinction.

within it are destroyed. Mount St. Helens is a volcano in the state of Washington. It erupted in 1980, destroying almost 100,000 acres of forest in just a few days. Over 7,000 elk, deer, and bears were killed. Fortunately, the area around the volcano was not the only place where these species lived. Twenty-five years later, these species are returning.

Extinction of the Dinosaurs

Can a single, sudden change in the environment lead to extinction? Many scientists believe that a single event caused the extinction of the dinosaurs. (See Figure 9.5-3.) A large asteroid struck Earth 65 million years ago in what is now Mexico. An *asteroid* is a piece of rock moving through space. The impact of the collision sent enough material into the atmosphere to change the global climate. Dust and ash filled the atmosphere and blocked the heating rays of the sun. The sudden cooling led to the loss of the dinosaurs' habitat, and

they disappeared in what is called a *mass extinction*.

The Effect of Environmental Change on Humans

Both gradual and sudden changes to the environment affect humans, too. Floods, droughts, earthquakes, and storms can cause changes to an environment that greatly affect the people living there. Hurricane Katrina, for example, caused the flooding of New Orleans, making much of the city unlivable. A drought in Africa is threatening millions of people with starvation. Without the needed rainfall, plants, the beginning of every food chain, cannot grow.

The **ozone layer** is a natural part of the atmosphere that prevents some of the sun's harmful energy from reaching the Earth's surface. Scientists have observed a reduction in the amount of ozone, especially in the southern hemisphere. This depletion of the ozone in the atmosphere increases our

Figure 9.5-3. The dinosaurs pictured here were victims of a mass extinction.

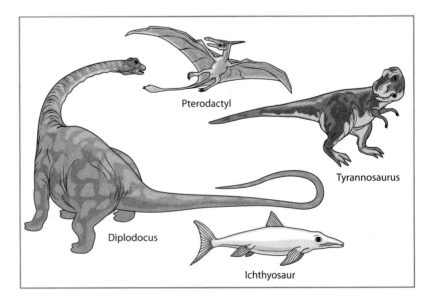

Pterodactyl

Tyrannosaurus

Diplodocus

Ichthyosaur

exposure to the sun's **ultraviolet rays**, which are high-energy rays. This increased exposure results in more severe sunburn, and an increase in the number of cases of skin cancer. It has become more important to wear sunscreen when you are going to be in the sun. One of the causes of ozone depletion was the use of chemicals called CFCs. These chemicals were used in refrigerators, air conditioners, and spray cans. We are no longer using these CFCs, and the ozone layer seems to be slowly recovering.

Interesting Facts About Ozone Depletion

Australia is located in the southern hemisphere and is greatly affected by the depletion of the ozone layer. To combat increasing rates of skin cancer, a campaign urging people to be "Sun Smart" was launched. A new saying in Australia is "Slip-Slop-Slap." Australians are asked to "slip" on a shirt, "slop" on some sunscreen, and "slap" on a hat.

Questions

1. Many forests of the southern United States have disappeared, and the population of ivory-billed woodpeckers has declined. Today, there may be only a few remaining. The cause of this decline is most likely
   (1) a contagious disease
   (2) loss of habitat
   (3) excessive hunting by humans
   (4) introduction of a predator

2. The complete and permanent loss of a species is called
   (1) predation   (3) extinction
   (2) parasitism   (4) migration

3. A sudden, drastic change in the environment can be the result of
   (1) a decrease in annual average rainfall
   (2) global warming
   (3) a volcanic eruption
   (4) global cooling

Thinking and Analyzing

1. Some naturalists claim to have seen an ivory-billed woodpecker in a forest in Arkansas. When John Yrizzary, a noted bird expert from Brooklyn, heard this news, he said, "Unless they find two of them, I'm not excited." Explain why Mr. Yrizzary feels this way.

Review Questions

Term Identification

Each question below shows two terms from Chapter 9. One of the terms is defined.
(1) Choose the term that matches the definition.
(2) Describe how the two terms are different. Following each term is the section (in parenthesis) where the description or definition of that term is found.

1. *Population (9.1) — Community (9.1)*
   All the members of a particular species that live within a given area

2. *Abiotic factors (9.1) — Biotic factors (9.1)*
   Nonliving parts of an environment

3. *Mutualism (9.3) — Commensalism (9.3)*
   A symbiotic relationship in which one species benefits, while the other is not affected

4. *Host (9.4) — Parasite (9.4)*
   The organism that is harmed from a parasitic relationship

5. *Ozone layer (9.5) — Ultraviolet rays (9.5)*
   High-energy radiation from the sun that has been linked to skin cancer

Multiple Choice (Part 1)

1. Which of the four terms below includes the other three terms?
   (1) community   (3) population
   (2) ecosystem   (4) individual

2. The drawing below shows a woodpecker using its long, sharp beak to capture and eat insects.

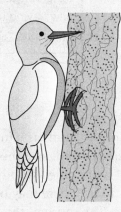

What factor might contribute to the extinction of this species of woodpecker?

(1) a new source of food
(2) an overabundance of trees
(3) the use of pesticides in the forest
(4) the increase in the population of insects

3. Which of the following statements best describes parasitism?
(1) It is a relationship in which both organisms benefit.
(2) It is a relationship in which both organisms are harmed.
(3) It is a relationship in which one organism benefits and the other is harmed.
(4) It is a relationship in which one organism benefits and the other is not affected.

Read the paragraph below to answer questions 4 and 5.

Lice are organisms that live among human hairs. They are commonly found in children between the ages of 3 and 12. They feed on very small amounts of blood taken from the human scalp. Lice do not spread disease but they are annoying. Their bite causes the scalp to itch. Often, scratching the bite can lead to irritation or infection. Doctors prescribe medicated shampoo to get rid of lice.

4. The relationship between lice and humans can best be described as
   (1) mutualism   (3) commensalism
   (2) parasitism   (4) communism

5. Humans are to lice as
   (1) hosts are to parasites
   (2) producers are to consumers
   (3) predators are to prey
   (4) pioneers are to succession

6. All the organisms of a particular species that live in an ecosystem make up the
   (1) environment   (3) biome
   (2) population   (4) community

Base your answers to questions 7 and 8 on the information and accompanying graph above:

In 1900, a river in Africa dried up. The dry riverbed allowed a pride of lions to migrate into a new environment that previously did not contain lions. The graph above shows how the lion population changed over time.

7. How many lions were in the original pride that migrated to the new territory?
   (1) 60 (2) 10 (3) 1900 (4) 30

8. Based on this graph, the carrying capacity of this environment for lions is closest to
   (1) 60 (2) 10 (3) 1900 (4) 30

9. The Galapagos Islands are known for unique species of organisms. One of these

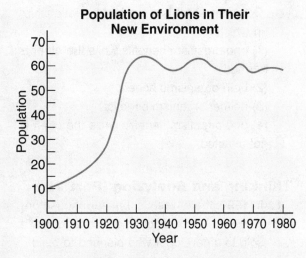

Population of Lions in Their New Environment

organisms is a bird called the Nazca Booby (shown below). The sharp-beaked ground finch, another bird unique to the Galapagos Islands, feeds on a parasite living in the feathers of the Nazca Booby. The relationship between the Nazca Booby and the sharp-beaked ground finch can best be described as
   (1) parasitism   (3) mutualism
   (2) commensalism   (4) predation

10. Commensalism is a symbiotic relationship in which

(1) one organism benefits while the other is harmed

(2) both organisms benefit

(3) neither organism benefits

(4) one organism benefits while the other is not affected

Thinking and Analyzing (Part 2)

1. In 1990, there was just one farmhouse on a square kilometer of land. The land was sold to a developer who planned to build more houses for a total of 20 houses. While some of the houses were being built, a scientist collected the data shown in the table below.

Year	Number of Rabbits per km²	Number of Houses per km²
1990	75	1
1991	72	2
1992	64	3
1993	60	5
1994	56	8
1995	51	10
1996	46	12
1997	35	13
1998	29	15
1999	21	16
2000	?	?

(a) The last four houses were built in the year 2000. Based on the data in the table, predict what happened to the number of rabbits in the year 2000.

(b) Explain how you arrived at this prediction.

2. Describe the differences between a population and a community.

3. Mutualism, commensalism, and parasitism are three basic types of symbiotic relationships. Describe how these symbiotic relationships are alike and how they are different.

4. Sometimes environments change very suddenly, while at other times environments change very slowly, perhaps taking thousands of years. Give an example of a sudden environmental change and a slow environmental change.

Chapter Puzzle (*Hint:* The words in this puzzle are terms used in the chapter.)

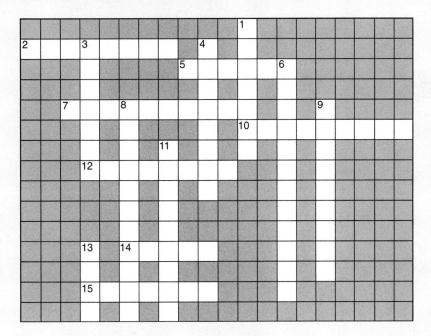

Across

2 an organism that harms another organism in a symbiotic relationship

5 the living parts of the environment are called the _____ factors

7 all the members of a particular species in a given area

10 all the different populations living in an area

12 the _____ capacity is the number of individuals that an ecosystem can support

14 the ___ layer protects Earth from ultraviolet rays

15 organisms of the same kind

Down

1 a species that no longer exists

3 nonliving

4 conditions that restrict the size of a population are called _____factors

6 a relationship in which one species benefits while the other is not affected

8 radiation from the sun that is linked to skin cancer

9 a symbiotic relationship in which both species benefit

11 a relationship between different species living in close permanent association

13 the organism harmed by a parasite

Glossary

A

abiotic (AY-by-AH-tik) **factors:** nonliving parts of an environment

adaptation (ah-dap-TAY-shun): a special characteristic that enables an organism to survive under a given set of conditions

air mass: a large body of air that has generally the same temperature and humidity conditions throughout

air pressure: the force caused by air pushing down on an object; air pressure is exerted in all directions

algae (AL-jee): a one-celled organism that contains chloroplasts and cell walls

aneroid (A-nuh-royd) **barometer:** a type of barometer that does not contain a liquid.

animals: a kingdom of multicellular organisms whose cells contain centrioles

aquatic (uh-KWAH-tic) **succession** (suhk-SEH-shun): ecological succession that begins with a lake or pond community

Archimedes' (ark-uh-MEE-deez) **principle:** the weight lost by an object placed in a fluid is equal to the weight of the fluid that the object displaces

B

bar graph: a graph that uses length to show differences between experimental groups

barometer (buh-RAHM-uht-uhr): an instrument used to measure air pressure

behavioral adaptation: activities performed by an organism that help it to survive under a given set of conditions

biotic (by-AH-tik) **factors:** living parts of an environment

biome (BUY-ohm): a region of Earth that is characterized by a particular climate, soil, and community of plants

boiling: a rapid phase change from liquid to gas that occurs at a particular temperature

boiling point: the temperature at which a liquid bubbles and turns to a gas

brood: a group of birds hatched at the same time and cared for by the same mother

buoyancy (BOY-uhn-see): the ability of an object to float in a fluid (liquid or gas)

buoyant (BOY-uhnt) **force:** the upward force acting on an object placed in a fluid

C

calorie (KA-luh-ree): the amount of heat energy needed to raise the temperature of 1 gram of water 1°C

carbon cycle: the process through which plants and animals exchange carbon in the form of sugar and carbon dioxide

carnivore (KAR-nih-vore): a meat-eating animal

carrying capacity: the number of individuals of a given species that can be supported by an ecosystem

cell: the building block of living things

chlorophyll (KLOR-uh-fill): a green chemical in plants needed for photosynthesis

cilia (SIH-lee-uh): tiny hairlike structures used for locomotion

classification: organization according to similarities

climate: the average weather in an area

climax (KLI-macks) **community:** the final community that appears and is not replaced through ecological succession

cold-blooded: an organism that does not maintain a constant body temperature

cold front: the boundary produced when cold air pushes under warm air

community: all the different populations that live within a given area

commensalism (koh-MEN-suh-liz-uhm): a symbiotic relationship in which one species benefits, while the other is not affected

competition: the interaction between organisms that require the same food and resources

complex machine: a machine made of two or more simple machines

computer model: a model generated by a computer program

condensation (kahn-duhn-SAY-shuhn): a phase change of a gas into a liquid

condensation (kahn-duhn-SAY-shuhn) **level:** the altitude where the temperature is low enough to cause water vapor to condense into tiny water droplets and to form clouds

condensing (kahn-DEN-sing): changing from gas to liquid

conduction (kuhn-DUHK-shuhn): the transfer of heat from one object to another object by direct contact

consumer (kuhn-SUE-muhr): an organism that obtains nutrients by eating other organisms

continental (kahn-the-NEN-till) **air mass:** a dry air mass that forms over land

contraction (kahn-TRAK-shuhn): a decrease in the length or volume of a substance

control or **control group:** a standard to which the experimental group is compared

convection (kuhn-VEHK-shuhn): the process of heat transfer within a liquid or gas

convection current: the movement of a liquid or gas caused by a change in temperature

cumulus (KYOO-myuh-luhs): puffy clouds

cumulonimbus (KYOO-myuh-low-NIM-buhs): a thunderhead; a cloud associated with a thunderstorm

cycle: a periodically repeated sequence of events

cytoplasm (SYE–toe-plaz-uhm): the gel-like substance that fills the cell

D

data: information

deciduous (dee-SID-you-us): a type of tree or shrub that drops its leaves seasonally

decomposer (dee-kuhm-POH-zuhr): special organisms, such as fungi (mushrooms and molds) and some bacteria, that break down dead animals' remains and return their nutrients to the soil

density (DEHN-sih-tee): a quantity that compares the mass of an object to its volume, determined by how close the molecules are to each other

dependent variable: the condition in an experiment that changes due to the effects of the independent variable

deposition (dee-PAH-zeh-shuhn): the phase change from gas to solid

desert (DEH-zurt): a biome that receives less than 25 centimeters (10 inches) of rain per year

dormancy (door-MUHN-see): a state in which an organism is inactive while it awaits more favorable environmental conditions

drought (DROWT): a lower-than-normal amount of precipitation for extended periods of time

E

ecological succession: the natural process by which one community of living things is replaced by another community, until a stable climax community appears

ecology: the study of the interaction between organisms and their environment

ecosystem: all the living and nonliving things that interact within a certain area

efficiency (e-FI-shan-see): the ratio of work output to work input

effort: the force applied to a machine

elevation (eh-luh-VAY-shun): the height above sea level

energy: the ability to do work

energy pyramid: a device used to show the energy relationships between producers and consumers in a food chain; the producers are always at the base of an energy pyramid

environment (en-VY-run-ment): the surroundings in which an organism lives, including both living and nonliving things

equator (ee-KWAY-tuhr): an imaginary line around the center of Earth midway between the North and South Poles

evaporating: a phase change from liquid to gas that occurs at the surface of a liquid at any temperature

evaporation (ee-vap-uh-RAY-shuhn): a phase change of a liquid into a gas

expansion (ik-SPAN-shuhn): an increase in the length or volume of a substance

experiment: an organized procedure to test an idea or gather new information

experimental group: a group in an experiment that receives the independent variable

extinct (ehks-TEENKT): species that no longer exist

F

flagellum (fluh-JELL-uhm): a tail-like structure used for locomotion

flood: the overflowing of water over land that is not usually under water

fluid: a material that flows; a liquid or a gas

food chain: a sequence of organisms through which energy is passed along in an ecosystem

food web: a number of interconnected food chains

fossil fuel: an energy source that is formed from the remains of plants or animals

freezing: a phase change from liquid to solid

freezing point: the temperature at which a liquid turns to a solid

friction (FRICK-shun): a force that resists the motion of an object when one surface moves over another surface

front: the boundary between different air masses

fulcrum (FUHL-krum): the point around which a lever turns

fungi (FUN-jye): kingdom of organisms whose cells contain cell walls but no chloroplasts

G

gas: the phase of matter that has no definite shape and no definite volume

global warming: a rise in worldwide temperatures

glucose (GLEW-kose): a simple sugar produced during photosynthesis

graduated cylinder (GRAJ-you-ay-ted SILL-in-duhr): a tool used to measure the volume of a liquid

gram: a unit of mass

greenhouse effect: a process that allows short-wave radiation to pass through glass or the atmosphere, but prevents long-wave radiation from passing back out; a heating of the atmosphere due to gases that trap and hold heat

H

habitat: the particular environment in which organisms live, such as a desert, a pond, a forest, or a jungle

hail: a form of precipitation formed by layer upon layer of ice; it is commonly round and pea size, but can sometimes be very large and irregularly shaped

heat: a form of energy produced by vibrating molecules in a substance

heating curve: a graph used to determine the melting point or boiling point of a substance

herbivore (UR-bih-vore): a plant-eating animal

hibernation (HI-buhr-NAY-shun): a sleeplike state of reduced body activity

high-pressure system: cooler, denser, sinking air that usually produces clear skies and fair weather

host: the organism that is harmed from a parasitic relationship

humidity (hyoo-MIHD-ih-tee): the amount of water vapor in the air

hurricane (HUR-ih-kayn): a large tropical storm with rotating wind speeds greater than 74 mph

hydrosphere (HY-droh-sfeer): the surface and subsurface water on Earth

hypothesis (hy-PAH-thuh-sis): a possible answer to a problem, based on observation and research

I

independent variable: a condition in an experiment that is deliberately changed by the scientist

infrared (in-fruh-RED) **radiation:** a type of radiation that transfers heat through space

inquiry: a system of scientific investigation

J

joule (JOOL): the amount of work done by a force of one newton over a distance of one meter; one joule equals one N-m

K

kilogram (KILL-oh-gram): a unit of mass, equal to 1000 grams

kinetic (kih-NEH-tik) **energy**: the energy an object has because it is moving

kinetic (kih-NEHT-ihk) **molecular** (muh-LE-kyuh-ler) **theory**: a theory that describes matter as containing molecules, with the molecules in constant motion (vibrating)

kingdom: the most general category of living things

L

latitude (LAT-ih-tude): a measure of distance north or south of the Equator, expressed in degrees

Law of Conservation (kon-sur-VA-shan) **of Energy**: states that energy in the universe cannot be created or destroyed; it can only be changed into another form

Law of Conservation of Matter: states that matter in the universe cannot be created or destroyed

life functions: processes that occur in all living things

limiting factors: conditions that limit the size of a population within an environment

line graph: a graph that connects numerical data points

liquid: the phase of matter that has no definite shape but has a definite volume

liter (LEE-tuhr): a unit of volume

lithosphere (LIHTH-oh-sfeer): the solid rock part of Earth

low-pressure system: warmer, less dense, rising air that usually produces cloudy skies and rainy weather

M

maritime (MEHR-eh-time) **air mass**: a moist air mass that forms over a large body of water

mass: a measure of the amount of matter in a sample

matter: anything that has mass and takes up space

maximum humidity: the greatest amount of water vapor the air can hold based upon the temperature of the air

mechanical (muh-KA-nih-kuhl) **energy**: the energy of a moving object, it is kinetic energy (KE)

melting: a phase change from solid to liquid

melting point: the temperature at which a solid turns to liquid

meniscus (muh-NIS-kus): the curve in the surface when a liquid, such as water, is placed in a glass container

mesosphere (MEHZ-uh-sfeer): third highest layer in the atmosphere where a decrease in temperature occurs with an increase in altitude

migration (my-GRAY-shun): seasonal movement of animals from one location to another

millibar (MIH-leh-bar): unit of air pressure shown on weather maps; one inch of mercury equals 33.9 millibars of air pressure

model: a representation of an object or system

molecule (MAH-luh-kyool): the smallest particle of a substance that has the properties of that substance

monera (muh-NAIR-uh): a kingdom of unicellular organisms whose cells do not have a nucleus—such as bacteria

multicellular (MUHL-tee-SELL-you-luhr): containing more than one cell

mutualism (MEW-chew-uhl-lis-uhm): a symbiotic relationship in which both species benefit

N

newton: a unit of force in the metric system (about 0.22 lb, or 1 lb = about 4.5 newtons)

newton-meter (N-m): a force of one newton acting over a distance of one meter

nucleus (NEW-clee-uhs): the structure within the cell that controls cell activities

O

observation: anything we perceive through one or more of our senses

occluded (uh-KLOOD-ed) **front**: the boundary

produced when a cold air mass pushes cool air up and over warm air

ocean current: flowing rivers of water in the ocean

omnivore (AHM-nih-vore): an animal that eats both plants and animals

organelle (or-guh-NEHL): a small structure within the cytoplasm that carries out a specific life function

organism: a living thing

ozone (OH-zone) **layer:** a region in the atmosphere that contains a large concentration of ozone gas, which blocks most ultraviolet radiation from the sun

P

parasite (PAH-rah-site): the organism that benefits from a parasitic relationship

parasitism (PAH-ruh-sih-tiz-uhm): a symbiotic relationship in which one species benefits, while the other is harmed

phase: (fayz) the physical form of matter, such as gas, liquid, or solid; also called the state

phase change: changing from one phase of matter to another phase of matter; for example, in *melting*, a substance changes from a solid to a liquid

photosynthesis (foe-toe-SIN-thuh-sis): the chemical process in which a green plant uses sunlight to convert carbon dioxide and water into glucose and oxygen

physical adaptations: body structures that help an organism survive under a given set of conditions

physical model: a model that you can see or touch

pie chart: a graph that compares the parts of a whole by representing them as slices in a pie

pioneer (pih-oh-NEAR) **species:** the first organisms to appear in a barren environment

plants: a kingdom of multicellular organisms whose cells contain chloroplasts and cell walls

polar air mass: a cool or cold air mass that forms north of the United States

population (pa-pew-LAY-shun): all the members of a particular species that live within a given area

potential (puh-TEN-shuhl) **energy:** the stored energy an object has because of its position or condition

precipitation (pree-sihp-uh-TAY-shuhn): liquid or solid water that falls to Earth's surface from the atmosphere; rain, snow, sleet, and hail are examples of precipitation

predator: (PREH-dih-tuhr): an animal, such as the lion, that must hunt and kill for food

prey: an animal the predator (for example, the lion) hunts, such as the zebra

primary consumer: an animal that eats producers; a herbivore

primary succession: the process of ecological succession that begins from barren rock

problem: the question to be investigated

producer: an organism that makes its own food; most producers are green plants

protists (PROE-tists): a kingdom of mostly unicellular organisms whose cells have a nucleus

protozoa (PRO-tuh-ZOH-uh): a one-celled organism that does not have chloroplasts or cell walls

pseudopod (SOO-doh-pod): an extension of cytoplasm used by an ameba for locomotion

psychrometer (sy-KRAHM-uh-tuhr): an instrument that uses a wet-bulb and a dry-bulb thermometer to measure relative humidity

Q

quantitative (KWAN-tih-tay-tiv): involving measured values

R

radiant (RAY-dee-uhnt) **energy:** energy, such as light, radio waves, and X-rays, that is capable of passing through space

radiation (ray-dee-AY-shuhn): the process of transferring energy in the form of waves

relative humidity: the ratio between the actual amount of water vapor in the air, and the maximum amount of water vapor the air can hold at a specific temperature

research: gathering information, facts, and the opinions of others

resistance (ri-ZIS-tans): the force a machine has to overcome

S

savannah (suh-VAH-nuh): a tropical grassland

scale model: a model in which the relative sizes and positions of all parts are maintained

scavenger (SKAH-ven-juhr): an animal, such as a vulture, that eats the remains of dead animals

scientific method: an organized, step-by-step approach to problem solving

secondary consumer: an animal that eats herbivores

secondary succession: the process of ecological succession that begins from preexisting soil

simulate: to imitate the functions of a system or process

sleet: a form of precipitation produced when rain freezes as it falls to Earth

solid: the phase of matter that has a definite shape and definite volume

species: a group of organisms of the same kind that produce offspring that can reproduce

starch: a nutrient produced from sugar that stores energy in plants

stationary (STAY-shah-nair-ee) **front:** the boundary produced when a cool air mass and a warm air mass meet and neither is capable of pushing the other

stratosphere (STRAT-uh-sfeer): second highest layer in the atmosphere where the temperature increases with an increase in altitude

stratus (STRAYT-uhs): flat, layered clouds

storm surge: a wall of ocean water pushed toward land during a hurricane

sublimation (sub-LY-may-shuhn): a phase change from solid to gas

substance: a pure form of matter that contains only one type of molecule

symbiosis (sim-by-OH-sis): a relationship between different species living in close, permanent association

T

taiga (TAY-guh): a biome with long, cold winters that contains needle-leaf evergreen trees, such as fir and spruce

temperate (TEHM-puhr-iht): a climate with warm summers and cold winters

temperate deciduous forest: a biome containing deciduous trees

temperate grassland: a biome that contains tall grass and few trees

temperature: a measure of the average molecular motion of the molecules of a substance; the greater the average molecular motion, the greater the temperature.

theory: a detailed explanation of the natural world based on the best information available

thermosphere (THUR-muh-sfeer): uppermost layer in the atmosphere where the temperature increases with an increase in altitude

thunderstorm: a storm with lightning and thunder, very gusty winds, heavy rain, and sometimes hail

tornado (tawr-NAY-doh): a violently spinning funnel-shaped cloud

tornado alley: an area, from Texas to Indiana, where a high number of tornadoes occur

transformation (trans-fur-MA-shan) **of energy:** the changing of one form of energy into another form

triple beam balance: a tool used to measure mass

tropical: a hot and moist climate

tropical air mass: a warm or hot air mass that forms south of the United States.

tropical depression: a storm formed over warm water with rotating wind speeds less than 39 mph

tropical rain forest: a biome with a warm, wet climate in which the climax community contains broadleaf evergreens

tropical storm: a storm formed over warm water with rotating wind speeds between 39 and 74 mph

troposphere (TROH-puh-sfeer): lowest layer of the atmosphere where the temperature decreases with an increase in altitude

tundra (TUHN-druh): a treeless biome with soil that is always frozen

U

ultraviolet rays: high-energy radiation from the sun that has been linked to skin cancer

unicellular (YOU-nih-SELL-you-luhr): consisting of a single cell

V

variable: a condition that changes during an experiment

volume: the amount of space that a sample of matter takes up

W

warm-blooded: an organism that maintains a constant body temperature

warm front: the boundary produced when warm air pushes up and over cooler air

water cycle: the process by which water moves back and forth between Earth's surface and the atmosphere by means of evaporation, condensation, and precipitation

water displacement (dis-PLACE-ment): the replacement of a volume of water by the same volume of another substance that is placed in it

water vapor: the gaseous form of water in the air

wavelength: the distance from the top of one wave to the top of the next wave

weather: the state of the atmosphere at a given location over a short period of time

work: what is done when a force moves an object a specific distance

X

x-axis: the horizontal axis of a graph on which the independent variable is shown

Y

y-axis: the vertical axis of a graph on which the dependent variable is shown

Index

Photo Credits